BOTANY AND BREEDING OF SPICE CROPS

a division of

NIPA GENX ELECTRONIC RESOURCES & SOLUTIONS P. LTD.

New Delhi-110 034

ABOUT THE AUTHORS

Dr. A.B. Sharangi is one of the Professors of eminence in Horticulture and former Head, Department of Plantation, Spices, Medicinal and Crops, Faculty of Horticulture, Bidhan Chandra Krishi Viswavidyalaya (Agricultural University) India. He is in the profession of teaching for twenty three years. He spent times in Prof. Cousen's lab in Melbourne, Prof Picha's lab in USA, Dr Dobson's lab in UK and continued research on herbs and spices. One of his papers has ranked among the top 25 articles in Science Direct. He has published about 90 research papers in peer-reviewed journal, 50 conference papers, 25 books from reputed publishers including Springer Nature as well as several book chapters published from Springer, Taylor and Francis, CRC Press, Nova Publishers (USA) and scores of popular scientific articles. Presently he is associated with 50 international and national journals as editor in chief, associate editor, technical editor, editorial board member and reviewer. Prof Sharangi has visited abroad extensively on academic mission and obtained several international awards viz., ENDEAVOUR Post-doctoral Award (Australia), INSA-RSE Visiting Scientist (UK), FULBRIGHT Visiting Faculty (USA), Achiever's Award (SADHNA), Marquis Who's Who (USA), Man of the Year-2015 (Cambridge, UK), Outstanding Scientist and Higher Education Leadership Award (Venus International Foundation), Bharat Ratna Mother Teresa Gold Medal Award-2020 (GEPRA), etc. He had a couple of invited lectures in UK, USA, Australia, Thailand, Israel and Bangladesh on several aspects of herbs and spices and associated with a number of research projects as Principal and Co-Principal Investigators having academic and empirical implications. He is an elected FELLOW of WB Academy of Science and Technology, (FAScT). He is also the Fellow of International Society for Research and Development (FISRD, UK), the Society of Applied Biotechnology (FSAB), Academy of Environment and Life Sciences (FAELS), International Scientific Research Organization for Science, Engineering and Technology (FISROSET), Scientific Society of Advanced Research and Social Change (FSSARSC), Society of Pharmacognosy and Phytochemistry (FSPP), International Society of Noni Sciences (FISNS). He is an active member of several science academies and societies like NASI, NABS, ISNS, IAHS (Formerly HSI), CWSS including the New York Academy of Science (NYAS), World Academy of Science, Engineering and Technology (WASET), African Forest Forum (AFF), to name a few.

Dr. Md. Nasim Ali is presently working as Associate Professor in the Department of Agricultural Biotechnology, Faculty of Agriculture, Bidhan Chandra Krishi Viswavidyalaya, Mohanpur, Nadia. He completed his PhD. (Ag.) in Genetics from Bidhan Chandra Krishi Viswavidyalaya, Mohanpur, Nadia, W.B. in 2007. After completion of PhD, he served as a Junior Scientist in Plant Tissue Culture & Biotechnology cell in West Bengal State Council of Science & Technology then worked for more than 8 years as Assistant Professor in the Department of Agricultural Biotechnology in the Integrated Rural Development & Management Faculty Centre of Ramakrishna Mission Vivekananda University, at Ramakrishna Mission Ashrama, Narendrapur, Kolkata. His research area may be bracketed as "Utilization of molecular marker and plant tissue culture technique in Crop Improvement". He is attached with 7 different national and state level funded research projects which included the funding agencies like NTRF, Kolkata, DST (central and State), BARC, Mumbai. He has also served as the coordinator of the DBT sponsored M. Sc. Teaching Programme at BCKV. He has been recognized by International Accreditation Organization (USA) and awarded Certified Faculty Member in 2013. He is also the recipient of CWSS Young Scientist Award and Fellow Award. He got 7 Best Paper award on different National/ International level Platforms. He is the honourable Life member of different scientific societies namely,The Indian Science Congress Association (ISCA), Association for Plant Breeding and Improvement (APBI), The Indian Natural Fibre Society (TINFS), The Crop and Weed Science Society (CWSS), Society for Application of Statistics in Agriculture and Allied Sciences (SASAA), Indian Society of Genetics and Plant Breeding (ISGPB). He is presently serving as the Editor of the Journal of Crop and Weed and also acts as reviewer/editorial board member of several internationally reputed scientific journals of his field. He attended more than thirty International or national seminar and workshops. Several times he acted as the resource person in the several training programmes organized by UNICEF in abroad and continuously acting as resource persons for many State/Central Universities and private and public sponsored organizations. Till date he has guided 3 Ph.D and 6 M. Sc students who have successfully completed their degrees. He has published more than 50 research articles, 2 books.

Dr. Syandan Sinha Ray completed his Ph.D in Agricultural Biotechnology from Ramakrishna Mission Vivekananda Educational and Research Institute (Formerly known as Ramakrishna Mission Vivekananda University) and received his Ph.D award on 2017. His research area covered the micro-propagation and macro-propagation economically important crops especially bamboo species. Till date he has published more than 10 research papers including research article, review article, and book chapters in various well known journals. He has also qualified ICAR NET in the Agricultural Biotechnology. He is also a life member of Crop and Weed Science Society (CWSS). Currently he is working as Teaching cum Lab Assistant at IRDM Faculty centre of Ramakrishna Mission Vivekananda Educational and Research Institute. As part of teaching, he teaches several aspects of Agricultural Biotechnology i.e. Plant tissue culture, Genetic engineering and Molecular tools and technique at Post Graduate level.

BOTANY AND BREEDING OF SPICE CROPS

A.B. Sharangi
Md. Nasim Ali
Syandan Sinha Ray

a division of
NIPA GENX ELECTRONIC RESOURCES & SOLUTIONS P. LTD.
New Delhi-110 034

a division of
NIPA GENX ELECTRONIC RESOURCES & SOLUTIONS P. LTD.
101,103, Vikas Surya Plaza, CU Block
L.S.C.Market, Pitam Pura, New Delhi-110 034
Ph : +91 11 27341616, 27341717, 27341718
E-mail:newindiapublishingagency@gmail.com
www: www.nipabooks.com
For customer assistance, please contact
Phone: + 91-11-27 34 17 17 Fax: + 91-11- 27 34 16 16
E-Mail: feedbacks@nipabooks.com

ISBN: 978-93-90591-01-5

Composed and Designed by NIPA.

KERALA AGRICULTURAL UNIVERSITY

Main Campus, Kerala Agril. University P.O. Thrissur - 680 656, Kerala
Phone: (R) 0487-2373017 (M) 9446513017
E-mail: kvptr@yahoo.com; peter.kv@gmail.com

Professor K.V. Peter
Ph.D., FNAAS, FNASc FNABS. FISVeg, FISGen., FHSI, H.FCHAI, FISNS
Former : Vice-Chancellor Kerala Agricultural University and
Director ICAR IISR Calicut

Foreword

India has an ancient and fascinating history of spices. In fact, India was known for spices-black pepper (king of spices) and cardamom (queen of spices). As per Indian Spices Act, there are 53 spices originated and grown in India. International Standard Organization (ISO) recognizes 109 spices. Foreign traders from China and Europe came to India in search of spices especially black pepper and cardamom grown primarily in the western ghats passing through erstwhile Malabar, Mysore and Madras provinces. Large cardamom is grown in the present Sikkim and Darjeeling of West Bengal. India enjoys the position of being the largest producer, consumer and exporter in the world for several consecutive years. Skill and technological innovations in value addition to spices in the form of oils, oleoresins, pigments and lipids are monopolized by Indian private sectors. In India, spices are grown in over a million hectare with annual production of 7 lakh tones. Almost all the states in India grow one or more spices-saffron, the costliest grown in Jammu and Kashmir to curry leaf, the cheapest grown throughout India. To enhance production, productivity and quality of spice crops with special reference to export and domestic demand, appropriate production technology is of paramount importance. The survey, collection, cataloguing and conservation of their gene pool also assume significance and a holistic and thorough understanding of their botany are a pre-requisite and essentiality as well. Besides, documentation of breeding behaviour and methodology using front line scientific technology(s) is extremely vital to make any development programme vibrant and dynamic. Digital technology has come in the most scientific way in documentation including retrieval. For example, most of the spices being either grown under rainfed production system or in limited water conditions, breeding and selection of drought resistant/tolerant genotypes of seed spices offer the best long term solution to minimize the risk and adverse impact of limited water availability especially under rainfed production system. The present "Botany and Breeding of Spice Crops" is an excellent compilation of morphology of spice crops, centres of origin, modes of reproduction, pollination and emasculation techniques, methods of classical breeding, next generation plant breeding including the theories and applications of various omics viz., metabolomics, genomics, transcriptomics, proteomics, etc and is expected to meet the long felt demand of a book on spices dealing with botany and breeding towards a holistic learning. Prof. Dr. A.B. Sharangi, the well-known academician and Former Head Department of Plantation, Spices and Medicinal Crops, Faculty of Horticulture, BCKVV, West Bengal has authored the book. Co-authors are Dr. Md. Nasim Ali, BCKVV and Dr. S.S. Ray, RKMVERI. The New India Publishing Agency, New Delhi is the publisher. The book will be an asset to students, teachers and farmers as well.

(K V Peter)

Preface

Spices are high value low volume unique food adjunct offering aroma, flavour taste, colour and palatability to our day to day food items. They provide minerals vitamins, antioxidants and many other useful secondary metabolites which are offering priceless values to our basic health care and nourishment and pretty at par with other foods and/or food supplements. Precisely, spices are an integral part of food based lifestyle all over the world in general and India in particular in the context of cultural, religious as well as social aspects. There are quite a few literature available concerning different scientific approaches on how the genetic architectures are routinely altered in order to have a new genotype. But a holistic philosophy is somewhat lacking towards development of this crucial area mostly due to exhaustive understanding of the related branches like botany including floral biologyof individual crops. It is, therefore, felt worthwhile to provide a conceptual draft regarding improvement of these vital crops contributing hugely in our life and wellbeing. In this book, a number of such concepts have been represented towards a comprehensive understanding of the related theoretical and applied aspects.

The book throws some insights on the concepts of origin and domestication of spice crops. There is elaborate description regarding different modes of reproduction of plants along with crop specific techniques regarding emasculation and pollination techniques. Conventional plant breeding methods has categorically been explained as these concepts would help spice breeders to choose the appropriate breeding tool. It is evident that, breeders are often confused with restricted options whether to select a particular breeding method to be used as situation-specific unique tool or whether to use a combination of different tools. To disentangle this obvious crisis, there are substantial modern biotechnological interventions to help achieve objective-oriented goals. Hence, a detailed idea from tissue culture to genomics assisted breeding especially the theory and applications of various omics viz., metabolomics, genomics, transcriptomics, proteomics, etc., has been elaborated. Finally, different spice breeding information has been comprehensively documented in this book with some useful references to future directions of spice breeding throughout the world.We strongly believe that this book shall pave the way for new age researchers, faculties, plant breeders, policy makers and amateur readers towards theoretical and empirical studies as well.

We would like to thankfully acknowledge Mr. Shouvik Garai and Dr. Sandip Debnath for contributing to the initial drafts for some parts of the book.

A.B. Sharangi
Md. Nasim Ali
Syandan Sinha Ray

Contents

1

Introduction

Plant breeding is described as an art and science of changing the genetic architecture of plants with a goal towards betterment of mankind. Apart from this generalized concept, plant breeding is accomplished through inter-disciplinary mode assorted from simple selection in a variable plant-population having some desirable traits to the application of biotechnology and more. Worldwide, cultivars of different crops are routinely manufactured by Multinational and National level Industries, Government Institutions, Universities, Crop specific Organizations, professional plant breeders and even by the individuals such as gardeners and farmers.

Plant breeding is basically operated by the principles of Genetics where selections are practiced from a variable population and the selected variation needs to be inherited. Three categorized broad goals of plant breeding are 1) high yield 2) uplift in quality and 3) resistance against biotic and abiotic stress.

Spice crops are known to add nutritive value and increase the palatability as well as quality of food. Therapeutic uses of spices are also well recognised. The spice industry in India and trade has shown stunning progress over the last 5 years with 120% increase in revenue which is very close to touch $ 3 billion. East Asia is the major market for spices, followed by America and the European Union and the world spice trade has almost touched $ 17 billion by 2020. So, increased production is of paramount importance across the world and the potential is increasing. On the other hand, quality improvement is one of the primary objectives of spice crop breeding apart from achieving high yield. Cultivar and species diversity are the principal components of diversity of spices (Sasikumar *et al.* 1999; Krishnamoorthy *et al.,* 1997; Prasath and Venugopal, 2004). A good amount of variability is found in many important Indian spices like black pepper, cardamom, cinnamon, curry leaf, ginger, garcinia, tamarind and turmeric. A number of cultivars have been released followed by intra-germplasm evaluation and selection in spices. Contribution of recombination breeding is less in spice crops with regard to other field crops and a number of valuable varieties have been bred by clonal and seedling selection (Peter, 2008). Deployment of high yielding, water use efficient, climate resilient genotypes and technologies are also necessary.

The modern tools of science using several '*omics*' is helping to unravel many metabolic pathways that contribute to the intrinsic quality of spices and the mechanism of host-pathogen interactions. Increased precision, coupled with suitable breeding strategies, can enhance the yield and quality of black pepper and other suiatable spice crops by 3 fold in next 10 years by extensive deployment of superior genotypes in coffee and tea gardens of Tamil Nadu, Odisha, Assam and West Bengal in collaboration with planters, Spices Board, DASD (MIDH) and state departments (Krishna Kumar, 2015). The research efforts so far resulted in release of 72 improved varieties of spices. The conventional and/or biotechnological methods have been employed in evolving the high yielding varieties of spices. Varieties tolerant to biotic stresses (Table 1.1), abiotic stresses (Table 1.2) and varieties with high quality attributes (Table 1.3) are available in spices (Prasath *et al.*, 2018).

Table 1.1: Spice varieties tolerant to different biotic stresses

Crop	Biotic stress	Variety
Black pepper	Tolerant to root knot nematode	Pournami
	Tolerant to foot rot diseases/ Phytophthora foot rot	IISR Thevam & IISR Shakthi
Small cardamom	Tolerant to rhizome rot	IISR Avinash
	Resistant to Katte diseases	IISR Vijetha & Appangala 2
	Tolerant to stem borer and thrips	PV 2
	Tolerant to azhukal disease	ICRI 2
	Tolerant to rhizome rot	ICRI 3
Turmeric	Resistant to rhizome rot	Suguna, Sudarshana, Pant Peetabh
	Resistant to nematodes	IISR Pragati
	Resistant to scale insects	BSR 2
	Resistant to leaf blotch	IISR Kedaram
Ginger	Resistant to root knot nematode	IISR Mahima
	Tolerant to soft rot and bacterial wilt disease	Athira, Karthika

Table 1.2: Spices varieties tolerant to different abiotic stresses

Crop	Abiotic stress	Variety
Black pepper	Shade tolerant	Panniyur-2
	Suitability to high elevations	IISR Girimunda, IISR Malabar Excel
	Wider adaptability	Sreekara, Subhakara, Panniyur 7
Small cardamom	Drought tolerant	ICRI 5
	Early maturing	PV 1, ICRI 1
Turmeric	Tolerance to Salinity	Narendra Haldi 98
	Drought prone areas	Co 1, BSR 1
Ginger	Wide adaptability	IISR Varada, Suprabha
Coriander	ACr 1	Resistant to stem gall
	Suitable for saline and alkaline and drought prone areas	Co.2
	Suitable for rainfed areas, tolerant to wilt, drought condition, medium duration	Sindhu
	Suitable for rainfed crop or limited moisture condition	RCr 20
Cumin	GC 1	Resistant to cumin wilt

Table 1.3 Spices varieties released with high quality attributes

Crop	Quality attribute	Variety
Black pepper	High oil and oleoresin	Sreekara, Subhakara, Pallode-2
	High piperine	Panniyur-2, IISR Malabar Excel
	High dry recovery	IISR Shakthi
Small cardamom	High oil	PV 2, Mudigere 1, ICRI 1, Appangala 1
	High 1,8 cineole	ICRI 3, Appangala 2
	High dry recovery	PV 2
Turmeric	High curcumin	IISR Prathiba, IISR Pragati, Megha Turmeric
	Essential oil	Sudarshana, Suvarna
	High dry recovery	Roma, Suroma
Ginger	Essential oil	Himgiri, Aswathy

Table 1.4: Pre breeding achievements of spices

Achievement	Reference
Interspecific hybrids in Piper - to confer Phytophthora resistance - to confer pollu beetle resistance	Vanaja et al. (2007) Sasikumar et al. (1999b)
Interspecific hybrid in vanilla to impart disease resistance	Divakaran et al. (2006)
Polypoidy in black pepper	Nair & Ravindran (1992)
Genomics and transcriptomics approaches to identify and isolate candidate genes for - foot rot resistance in black pepper - bacterial wilt resistance in ginger	Neema & Johnson (2017) Prasath et al. (2014)
Mutation breeding of ginger to induce rhizome rot resistance	Prasath et al. (2015)

(Prasath *et al*., 2018)

In the conventional spice breeding, major emphasis has been placed on generating varieties which are fertilizer responsive, show resistance to various biotic and abiotic stresses and may be better in certain quality. However, when generating genotypes for the organic agriculture, emphasis will have to shift from the conventional approach to more targetted environment,locally adapted germplasm interactions between genotype and environment (GxE), and more importantly the participation of farmers in the selection process to exploit their huge knowledge of local varieties, appropriate production techniques and crop-environment interactions. Moreover, it is to be kept in mind that the spice varieties that have been specifically bred for the intensive agriculture may not perform well under the organic agriculture. This also needs altering the ideotype so that suitable varieties for organic cultivation can be developed (Sastry, 2015).

Climate resilient varieties of turmeric *viz.*, IISR Pragati (short duration,nematode resistant variety overcoming drought also) and NDH 98 (saline tolerant, stable yield across the country), ginger variety-IISR Mahima (nematode tolerant), cumin variety-GC4 (wilt tolerant), dual purpose varieties of coriander and determinate types of fenugreek (suitable for mechanized harvesting) played an imperative role in escalating the income of farmers. A few pre-breeding achievements in spices are given in Table 1.4. In many seasonal crops, early duration drought, heat and flood tolerant varieties and varieties for other adverse conditions are available, but in spice it is few and far between. It has to go a long way to breed suitable ideotype in spice to tackle climate change (Sastry, 2017).

2

Morphology of Spice Crops

A. General Morphology

2.1 The Root

It is the organ of a plant that typically lies below the surface of the soil.The root is best defined as the non-leaf, non-nodes bearing parts of the plant's body. However, important internal structural differences between stems and roots exist. Root spices are, rather surprisingly, rare. Some popular examples of spices having roots are: *Apium graveolens* (Celery), *Armoracia rusticana* (Horse radish), *Coriandrum sativum* (Coriander), *Ferula asafoetida* (Asafetida), *Glycyrrhiza glabra* (Licorice), *Levisticum officinale* (Lovage), (*Petroselinum crispum* (Parsley), etc. All of the mentioned plants have aromatic roots.Roots of gingiberaceae are adventitious and storage roots arise from among the nodes of these scales.

2.2 The Stem

The term "shoots" is often confused with "stems". "Shoots" generally refers to new fresh plant growth including both stems and other structures like leaves or flowers. In most plants, stems are located above the soil surface but some plants have underground stems. Stems have two pipe-like tissues called xylem and phloem. Pure stem spices are uncommon. Stems tend to be woody and unfriendly to eat, so the leaves are preferred. However, some subterranean fleshy stems known as rootstocks or rhizomes, form valuable spices, and, there are some aromatic barks as well. The axillary buds of ginger shoot up as leafy stem known as pseudo stem, which die out annually but the plant continues to live through its rhizome.

2.2.1 Rhizomes

Some popular examples of spices having rhizomes are: *Alpinia galanga* (Galangal), *Curcuma longa* (Turmeric), *Curcuma zedoaria* (Zedoary), *Houttuynia cordata* (Chameleon plant), *Kaempferia galanga* (Lesser Galangale), *Wasabia japonica* (Wasabi), *Zingiber officinale* (Ginger), etc.

2.2.2 Barks

Some popular examples of spices having rhizomes are: *Cinnamomum burmannii* (Indonesian Cinnamon) *Cinnamomum cassia* (Chinese Cinnamon, Cassia), *Cinnamomum loureiroi* (Vietnamese Cinnamon), *Cinnamomum zeylanicum* (Cinnamon), etc.

The following section will describe some specific spices where stem is an important consideration.

Cardamom : Thick, fleshy rhizomes and leafy stems, attaining a height of between 1.2 and 5 meters.

Turmeric : Stem : Rhizomes swollen, fleshy, copiously branched, creeping, subterraneans, emitting numerous, yellowish white or sordidly yellow roots. Main rhizome hard, ellipsoid, sordidly brown on the outside, with a spicy smell when bruised, within dark orange-coloured (bright carrot-coloured) with a somewhat lighter coloured margin or rind of 0.2-0.3 cm thickness, when newly cut somewhat oily-viscous to the touch, 5-7 cm long 2-3 cm diameter; lateral rhizomes numerous, much narrower, oblong or cylindrical, with numerous short lateral branches arranged in two rows, grayish brown on the outside, yellowish-orange coloured or vitelline within, 5-10cm long, 1-1.5 cm thick.

Ginger : Spurious stems closely approximate, erect or slightly drooping, leafy, 30-100 cm high.

Coriander : Erect, fistibular, herbaceous, branched, smooth.

Fennel: The stem is cylindrical, smooth and hollow at maturity with distinct veins. The branches also emerge later from the base of the plant.

Cinnamom: The stem reaches a height of 8-17 m in the wild. In an unharvested state, the trunk is stout, 30-60 cm in diameter, with a thick, grey bark and the branches set low down.

2.3 The Leaf

A leaf is an organ of a vascular plant and is the principal lateral appendage of the stem. A leaf is a dorsiventrally flattened organ, usually borne above ground and specialized for photosynthesis. Leaves are the most fragrant part. Bulbs are special subterranean reserve structures derived from leaves.

2.3.1 Leaves only

Popular examples of spices where only leaves have significance are: *Cinnamomum tamala* (Indian Bay-leaf), *Eugenia polyantha* (Indonesian Bay-leaf), *Laurus nobilis* (Bay leaf), *Murraya koenigii* (Curry-leaf), *Myrtus communis* (Myrtle), *Myrica gale* (Gale), *Pandanus amaryllifolius* (Screw pine leaf), *Peumus boldus* (Boldo leaves), *Piper auritum* (Mexican pepper-leaf), *Sassafras albidum* (Sassafras leaf), etc.

2.3.2 Both Leaves and stems

Popular examples of spices where both leaves as well as stems have significance are: *Allium schoenoprasum* (Chives), *Allium ursinum* (Bear's garlic), *Anethum graveolens* (Dill), *Anthriscus cerefolium* (Chervil), *Apium graveolens* (Celery), *Artemisia abrotanum* (Southernwood), *Artemisia dracunculus* (Tarragon), *Borago officinalis* (Borage), *Chenopodium ambrosioides* (Epazote), *Coriandrum sativum* (Coriander), *Cymbopogon citratus* (Lemon grass), *Eruca sativa* (Rocket), *Eryngium foetidum* (Long coriander), *Foeniculum vulgare* (Fennel), *Houttuynia cordata* (Chameleon plant), *Hyssopus officinalis* (Hyssop), *Lepidium sativum* (Cress), *Limnophila aromatica* (Rice paddy herb), *Levisticum officinale* (Lovage), *Maiorana hortensis* (Marjoram), *Melissa officinalis* (Lemon balm), *Mentha piperita* (Peppermint), *Myrrhis odorata* (Cicely), *Ocimum basilicum* (Basil), *Origanum vulgare* (Oregano), *Perilla frutescens* (Perilla), *Persicaria hydropiper* (Water pepper), *Persicaria odorata* (Vietnamese coriander), *Petroselinum crispum* (Parsley), *Rosmarinus officinalis* (Rosemary), *Ruta graveolens* (Rue), *Salvia officinalis* (Sage), *Satureja hortensis* (Savoury), *Spilanthes acmella* (Paracress), *Tagetes lucida* (Mexican tarragon), *Thymus vulgaris* (Thyme), *Trigonella caerulea* (Blue Fenugreek), etc.

Leaves of cinnamon are stiff, extipulate, opposite, somewhat variable in form and size. Petiole 1-2 cm long, grooved on the upper surface. Lamina usually 5-18 x 3-10 cm, ovate or elliptic; base more or less rounded and the tip tends to be somewhat acuminate. There are 3, sometimes 5, conspicuous longitudinal veins found at the base of the lamina and running almost to the tip. The young leaves of the flush are reddish, later turning dark green above with paler veins and pale glaucous beneath.

2.2.3 Bulbs

Popular examples of spices where bulbs are important are: *Allium cepa* (Onion) and *Allium sativum* (Garlic), etc.

The following section will describe some specific spices where leaf is an important consideration.

Black pepper: Leaves distichous, shortly stalked, broadly ovate-elliptic or ovate-oblong; base oblique, Obtuse or rounded, apex rather abruptly acuminate; coriaceous, above dark green, shining, beneath whitish green or sea-green, dull, densely, studded with white, pellucid dots, usually 5-7 nerved, rarely up to 9-nerved, 5-18 cm long, 2-12.5 cm wide ; petiole furrowed on the anterior side, sheating in the lower half 1-4 cm long.

Cardamom: The leaves are sub sessile, 30-60 cm long and 7.5 cm wide, oblong-lance late, and pubescent below.

Nutmeg: Distichous, petioles, rather small, elliptical-lanceolate or obovate; base acute, cuneate or obtuse; apex shortly narrowed, acuminate, acute, entire, coriaceous, above dark green, shining, densely stubbed with fine dots, beneath pale green, dull, soon glabrous on both surfaces, pin-natinerved, 5-15 cm long, 3-7cm wide, fragrant when bruised; petiole half-terete, glabrous, 0.7-1.5 cm long.

Turmeric : Leaf-sheaths conduplicate, with membranous margins, on both sides light green, longitudinally striate and finely grooved, dull, tapering towards the apex and rather abruptly passing into the petiole. Leaves distichous, alternate, obliquely erect, on shorter or longer petiole. Leaves distichous, alternate, obliquely erect, on shorter or longer petioles or subsessile, oblong-lanceolate; base cuneate, obtuse or almost rounded, decurrent into the petiole, apex narrowed, caudate, acute; above dark green with a green costa, feebly shinning, beneath very light green, on both sides glabrous, densely studded with pellucid dots especially beneath, 7-70 cm long, 2.5-18 cm wide, petiole broadly furrowed, along the margins with narrow, erect wings, light or dark green, densely dotte, 0.5-10 cm long, at the base with a ligule; ligule soon dotted, 0.5-10 cm long, at the base with a ligule; ligule soon withering, narrow, semi-annular, reflexed, ciliate, membranous.

Ginger : Leaves with a bug like smell when bruised, distichous, alternative, largest near the top of the stem, grass-like, very shortly stalked or subsessile; leaf-sheaths large, forming together a spurious stem, dark green, densely dotted, hairy at the top. Lamina linear-lanceolate; base obtuse or rounded, apex tapering, acute; above dark green with scattered white dots, shining, beneath light green, very densely studded with white dots, feebly shining, arachnoid or subglabrous, on an average 5-25 cm long, 1-3 cm wide; petiole hairy, broadly furrowed on the anterior side, at the base with a ligule; ligulebroad, short, membranous, white, appressed, with a truncate or faintly emarginated apex, 0.3-0.6 cm long.

Coriander : Ex-stipulate, alternate, lower leaflets are sessile, bifoliate or trifoliate. The lower leaves are broad and deeply segmented, the upper leaves are divided into linear lobes.

Fenugreek: Alternate, stipulate, pinnately trifoliate.

Fennel: The leaves are alternate, pinnately compound with partile leaf blades and sheathed petiole.

Garlic : Narrow flat leaves

Cumin : The leaves are pinnate or bi-pinnate with thread-like leaflets.

Saffron : Leaves 5–6, hysteranthos or with the tips just showing at anthesis, slightly grayish-green, 0.5–1 mm wide, glabrous.

Tamarind : The leaves are normally evergreen but may be shed briefly in very dry areas during the hot season

Dill : Dill's green leaves are wispy and fernlike and have a soft, sweet taste. The plant grows ordinarily from 2 to 2 1/2 feet high and is very like fennel, though smaller, having the same feathery leaves, which stand on sheathing foot-stalks, with linear and pointed leaflets.

Clove : The many branches of the clove tree are semi-erect with smooth oval shaped leaves. The branches end with a 3–4 flowers near the tip with one terminal flower and the others opening below it. The leaves, flowers and bark all have a distinct smell.

2.4 The Inflorescence

An inflorescence is a group or cluster of flowers arranged on a stem that is composed of a main branch or a complicated arrangement of branches. The following section will describe some specific spices where inflorescence is an important consideration.

Black pepper: Inflorescence is spike, spikes are filiform, curved or straight pendulous with sessile bracts bearing about 50-150 flowers per spike. The spike length depends on cultivar and influenced mostly by environment. Greenish yellow flowers are arranged in spikes along the spine. Anthesis starts from the base of the spike to tip over a period of 7-8 days. Wild forms of flower are usually dioecious but most cultivated ones are gynomonoecious. They do not have perianth, however are surrounded at their base by four bracts. Gynoecium (ovary) occupies the centre of the flower. 3-5 star shaped prominent stigmas and a very short style present. Androecium is composed of 2-4 stamens of approximately 1mm in length located on either side of ovary. The fruit is a berry, spherical, pungent and red when ripe and it appear 9 months after flowering. It is an oval drupaceous berry. In each spike there may be 50- 60 fruits each with a single seed.

Ginger: Inflorescence spiciform, produced direct from the rootstock ; peduncle leafless, erect, enclosed by green spathaceous scales, thin, glabrous, densely dotted, 15-25 cm diameter; bracts rather numerous, appressed, ovate-elliptical, with almost rounded or shortly cuspidate tips and thin, red or purple margins, green or yellowish green, afterwards yellowish, 2-3 cm long, 1.5-2 cm wide.

Flowers are borne on a spike produced in a peduncle different from the aerial leafy stem, arising directly from the rhizome. Flowers are many, trimerous bisexual, irregular, epigenous, yellow in colour with dark purplish spots. Androecium consists of stamens of which the outer 3 are reduced to stamenoids. The inner lateral stamens are united and showy to form a deep purple coloured labellum. The posterior stamen of theinner whorl is the only fertile stamen, which is enclosed by the labellum. The filament isflat and short with two prominent anther lobes. Anther cells are contiguous, produced into a long beak.

Ovary is inferior, three carpelled, three celled. Ovules are many on axial placentation. Style is long, delicate, lying in a groove in the stamen. Stigma is small and sub-globose.

Coriander: Compound umbel.

Fenugreek : Papilionaceous type

Cumin : The inflorescence is a compound umbel with white or pinkish flowers.

Fennel : The inflorescence is terminally born compound umbel subtended by involucres of bracts.

2.5 The Flower

Flower, the reproductive portion of any plant in the division Magnoliophyta (Angiospermae), a group commonly called flowering plants or angiosperms. As popularly used, the term "flower" especially applies when part or all of the reproductive structure is distinctive in colour and form.

Although most entomophilous flowers are somewhat fragrant, only few of them are used in the kitchen, because they usually lose their scent quickly when dried. The ripe form of flowers, fruits, are also frequently aromatic, but mostly not to attract insects but to repel them.

2.5.1 Buds

Popular examples of spices where buds have significance are: Capers (*Capparis spinosa*), Cloves (*Syzyium aromaticum*), etc.

2.5.2 Flowers or flower parts

Popular examples of spices where flowers or its parts have significance are: *Artemisia vulgaris* (Mugwort), *Carthamus tinctorius* (Safflower), *Crocus sativus* (Saffron), *Lavandula angustifolia* (Lavender), *Pandanus odoratissimus* (Kewra, Screw pine flower), *Rosa damascena* (Rose), etc.

The following section will describe some specific spices where flower is an important consideration.

Black pepper : Whitish to yellow-green small sessile flowers are arranged spirally along the spike. Flowers are bracteolate, with four peltate bracts. Bract of the female flowers broadly ovate, fleshy, adnate to the rachis

Cardamom : The flowers are borne in panicles that arise from the base of the vegetative shoots, upright at first but eventually becoming prostrate . The flower bracts are persistent, linear-oblong, up to 5 cm in length.

Turmeric : Flower long, narrow, white or yellowish white, 4.8-5.6 cm long. Bracteoles rather large , much longer than the corolla-tube, 3.5 cm long.

Ginger : Flowers in the axils of the bracteoles, ovate-oblong or narrowly elliptic, acute or obtuse, 2-2.5 cm long, 1-1.5 cm wide, almost as long as the bracts or slightly longer.

Coriander : Bisexual, complete, epigynous actinomorphic.

Fenugreek : Axillary, solitary cyme

Fennel : The inflorescence is terminally born compound umbel subtended by involucres of bracts. The number of umbels per plant varies widely depending upon the varity and agronomic conditions. The flowers are small, hermaphrodite, complete, regular and pentamerous.

Garlic : with and bears small white flowers and bulbils.

Vanilla : There may be up to 100 flowers on a single raceme, but usually no less than 20. The flowers are quite large and attractive with white, green, greenish yellow or cream colors. The flowers' sepals and petals are similar. The lip is tubular-shaped and surrounds the long, bristly column, opening up, as the bell of a trumpet, at its apex.

Cumin : The flowers are small and either pink or white colored. Typical to the Apiaceae family the flowers are born in umbels, and each umbel has 5 to 7 umbellets. The fruit is a schizocarp, 4-5 mm long, containing two mericarps with a single seed.

Saffron : Flowers autumnal, 1–3, white, occasionally white dark veins near the base of the segments, rarely very pale lilac; throat whitish or pale yellow, glabrous. Flowering October–November

Tamarind : The flower buds are distinctly pink due to the outer color of the 4 sepals which are shed when the flower opens. Inconspicuous, inch-wide flowers, borne in small racemes, are 5-petalled (2 reduced to bristles), yellow with orange or red streaks.

Cinnamon : The flowers, which are arranged in panicles, have a greenish colour and a distinct foetid odour. Flowers borne in lax axillary and terminal panicles on the ends of twigs. Peduncles creamy white, softly hairy, 5-7 cm long. Individual flowers are very small, about 3 mm in diameter, pale yellow, each subtended by a small, ovate, hairy bract. The calyx is campanulate and pubescent with 6 acutely pointed segments. Corolla absent.

Clove : Clove, *Syzygium aromaticum*, is a monoecious (both male and female flowers on the same plant) evergreen tree in the family Myrtaceae grown for its aromatic flowers.Cloves are the aromatic flower buds of a tree in the family Myrtaceae, Syzygiumaromaticum. The clove of commerce is the air-dried unopened flower bud obtained from evergreen medium sized tree.

2.6. The Fruit

Fruit, the fleshy or dry ripened ovary of a flowering plant, enclosing the seed or seeds.

2.6.1 Cones

Popular example of spices where cones have significance are: Juniper (*Juniperus communis*), etc

2.6.2 Pods

Popular examples of spices where pods have significance are: *Amomum subulatum* (Black Cardamom), *Capsicum annuum* (Bell pepper), *Capsicum frutescens* (Chile), *Elettaria cardamomum* (Cardamom), *Illicium verum* (Star anise), *Tamarindusindica* (Tamarind), *Vanilla planifolia* (Vanilla), *Xylopia aethiopica* (Negro pepper), *Zanthoxylum piperitum* (Sichuan pepper), etc.

2.6.3 Drupes and Berries

Popular examples of spices where drupes and berries have significance are: Laurel (*Laurus nobilis*), Cubeb pepper (*Piper cubeba*), Allspice (*Pimenta dioica*), Long pepper (*Piper longum*), Black pepper (*Piper nigrum*), Sumac (*Rhus coriaria*), Pink pepper (*Schinus molle*), Tasmanian pepper (*Tasmannia lanceolata*), Chaste tree (*Vitex agnus-castus*), etc.

2.6.4 Umbelliferae fruits

Popular examples of spices where drupes and berries have significance are: Dill (*Anethum graveolens*), Celery (*Apium graveolens*), Black Cumin (*Bunium persicum*), Caraway (*Carum carvi*), Coriander (*Coriandrum sativum*), Cumin (*Cuminum cyminum*), Fennel (*Foeniculum vulgare*), Lovage (*Levisticum officinale*), Anise (*Pimpinella anisum*), Ajwain (*Trachyspermum copticum*), etc.

The following section will describe some specific spices where fruit is an important consideration.

Black pepper : Fruiting spikes 2-20 cm long, berries sessile, globose, red when ripe, black when dry, 0.3-0.6 cm diam

Nutmeg : Fruit : Fruits globose-pyriform, shortly acuminate, 3-9 cm diameter, 2-valved when ripe, yellowish green on the outside, glabrous; pericarp fleshy, thick .

*Ginger:*Fruit is very rarely produced, which is an oblong capsule

Turmeric : Capsule, endospermous

Coriander : Globose, elliptical, cremocarp, which splits into two mericarp

Fenugreek : Pod

Fennel : The fruit, commonly known as seed, is a schizocarp of two mericarps attached to a dividing carpophores. A fully grown fruit is 4 to 10 mm long. Normally fruits are light greenish to grey in colour.

Vanilla: The fruit is termed "vanilla bean", though true beans are fabaceaen eudicots not at all closely related to orchids. Rather, the vanilla fruit is technically an elongate, fleshy and later dehiscent capsule 10-20 cm long. It ripens gradually for 8 to 9 months after flowering, eventually turning black in color and giving off a strong aroma. Each pod contains thousands of minute seeds, and both the pods and seeds within are used to create vanilla flavouring. Vanilla beans are harvested by hand from commercial orchards.

Tamarind : The fruits, flattish, bean like, irregularly curved and bulged pods, are borne in great abundance along the new branches and usually vary from 2 to 7 in long and from 3/4 to 1 1/4 in (2-3.2 cm) in diameter. Exceptionally large tamarinds have been found on individual trees. The pods may be cinnamon-brown or grayish-brown externally and, at first, are tender-skinned with green, highly acid flesh and soft, whitish, under-developed seeds. As they mature, the pods fill out somewhat and the juicy, acidulous pulp turns brown or reddish-brown. Thereafter, the skin becomes a brittle, easily-cracked shell and the pulp dehydrates naturally to a sticky paste enclosed by a few coarse strands of fiber extending lengthwise from the stalk. The 1 to 12 fully formed seeds are hard, glossy-brown, squarish in form, 1/8 to 1/2 in (1.1-1.25 cm) in diameter, and each is enclosed in a parchment like membrane.

Cumin : The fruits have eight ridges with oil canals. Seeds are hairy, in some varieties these hairs are prominent, and otherwise it is difficult to see them.

Cinnamon : The fruit is a purple 1-cm drupe containing a single seed.

Allspice : The fruits as found in commerce are small nearly globular berries, about 3/10 inch in diameter, somewhat like black pepper in appearance, with a rough and brittle surface and crowned by the remains of the calyx teeth, surrounding the short style. The fruit is two-celled, each cell containing a single, kidney-shaped seed. The remains of the the fruit of which is one-celled, one-seeded and grey and from Black Pepper corns, which are also one-celled and one-seeded calyx crowning the fruit and the presence of two single-seeded cells are features that distinguish Pimento from Cubebs.

Cinnamon: Fruit is a fleshy ovoid drupe, black, 1.5-2 cm long when ripe, with the enlarged calyx at the base.

2.7. The Seed

A seed is an embryonic plant enclosed in a protective outer covering. The formation of the seed is part of the process of reproduction in seed plants, the spermatophytes, including the gymnosperm and angiosperm plants. Popular examples of spices where seeds have significance are: Grains of paradise (*Aframomum melegueta*), Annatto seeds (*Bixa orellana*), Black mustard seed (*Brassica nigra*), Pumpkin seed (*Cucurbita pepo*), Tonka bean (*Dipteryx odorata*), Cardamom (*Elettaria cardamomum*), Nutmeg and Mace (*Myristica fragrans*), Onion seed (*Nigella sativa*), Poppy(*Papaver somniferum*), Almond (*Prunus dulcis*), Mahaleb cherry (*Prunus mahaleb*), Pomegranate seed (*Punica granatum*), Sesame (*Sesamum indicum*), White mustard seed (*Sinapis alba*), Fenugreek (*Trigonella foenum-graecum*),etc.

The following section will describe some specific spices where seed is an important consideration.

Cardamom : The oblong seed capsule is about 2.5 cm long and marked with fine vertical ribs.

Nutmeg: Seed broadly ovoid-oblong or subglobose, 1.5-4.5 cm long, 1-2.5cm thick brown on the outside, shinning; kernel whitish, much plaited and ruminate; aril rather thick, dark red, sometimes yellowish white, laciniate, enclosing the seed asanet .

Fenugreek: Small, greenish or brown

Chilli : Contains a very low amount of water but rich in lecithin. Contrary to popular belief, the seeds do not contain capsaicin, although since they are accanti to capsaiciniche glands, can absorb a little.

Dill : Dried dill seeds are light brown in colour and oval in shape, featuring one flat side and one convex ridged side. The seeds are similar in taste to caraway, featuring a flavour that is aromatic, sweet and citrusy, but also slightly bitter. Both seeds and leaves are valued as spice.

B. Detailed Botany and Floral Biology of Some Important Spice Crops

1. Black Pepper

Systematic position

Kingdom: Plantae

Subkingdom: Tracheobionta

Superphylum: Spermatophyta

Phylum: Magnoliophyta

Class: Magnoliopsida

Order: Piperales

Family: Piperaceae

Genus: *Piper*

Species: *P. nigrum*

Black pepper (*Piper nigrum L.,* Piperaceae) – 'King of spices' – is one of the oldest spices known. It originated in the humid, tropical evergreen forests of Western Ghats of India and is now grown in more than 25 countries, particularly in tropics. Important growing locations are India, Indonesia, Malaysia, Brazil, Thailand, Sri Lanka, Vietnam and China.

Cytology

Cytology of Piper has been studied by various workers and most of them are confined to the determination of chromosome number (Rahiman and Nair, 1986 and Samuel, 1986). The reported chromosome numbers 2n = 24, 26, 36, 39, 40, 48, 52, 60, 64, 65, 68, 78, 80, 96, 104, 132 etc. point to the existence of a polyploid series in the genus Piper. All the species studied from South India and Sri Lanka could be traced to a common basic number x = 13 while the North Indian species seem to have a basic number x = 12. The haploid number n = 12 as seen in *P. cubeba* may represent comparatively a primitive number from which x = 13 might have evolved (Jose and Sharma, 1985).

Botany

Piper nigrum is a climbing evergreen plant, growing to a height of 10 m or more. The vines branch horizontally from the nodes and do not attain much length. The pepper branches are dimorphic comprising of (i) the orthotropic vegetative climbing branches with 5-12 cm long internodes which give the framework of the plant, their stems are swollen at the nodes when young but become woody on aging; and (ii) a plagiotropic fruiting branch develops from the axillary bud present at each swollen node beside a leaf. The short adventitious roots produced at the nodes enable the plant to cling to the standard (climbing support). The leaves are alternate and simple, dark green and shiny above and pale green underneath. They are smooth and entire, broadly lanceolate but there is wide variation in leaf shape. The petioles are 2-5 cm long, and are grooved above.

1. Main stem : Originated from a seed or from stem cutting, it climbs on a support with the half of aerial or adventitious.
2. Runner shoots: Are produced from the basal portion of the main stem, growing at rate angle to the main stem, usually restricted upto 50 cm from the ground.
3. Fruiting branches (plagiotropes): Are produced from the nodes of the main stem and they grow laterally more or less to the right angle to main stem bears berries/spikes.
4. Top shoots (Orthotropes): After a period of growth, top portion of the main stem attain busty appearance with short thicker inter nodes and profuse branching with large number of adventitious roots at the nodes. This portion of the main stem/shoot called top shoot or othotropes. 5. Hanging shoots (Geotropes): In a fully grown vine at the top portion, some plagiotropes are give rise to a special type of shoots which hang down and grow geotropically. Leaves: Alternate, simple, lamina leathery, ovate, round or obtuse

Floral biology

The spike is a catkin type of inflorescence, which emerges opposite the upper leaves, on the plagiotropic branches. It is 3-15 cm long with 50-150 minute white to pale yellow flowers. The flowers are either unisexual with monoecious or dioecious forms, or bisexual, as seen in many cultivars. Under intense shade conditions, the hermaphrodite type produces more female flowers (Fig. 2.3) and less bisexual flowers (Fig. 2.4). The flowers are cross-pollinated and thus much variation occurs among the progeny. Moreover, hermaphrodite flowers are protogynous; i.e., they attain receptivity well in advance of the pollen maturity, which results in uneven fertilisation of flowers and non uniform, incomplete filling of the spikes.

Flowering begins at the base of the spike and continues to the tip over a week. It is found that bagged inflorescence produces fruits, indicating that the hermaphrodite cultivars (Eg: Balankotta, Kalluvalli) are self-fertile and self-pollination can occur without the aid of rain or wind. The pollen is in glutinous masses of several grains, which is broken up by light showers and the grains are trapped in the papillae of the stigma (Fig. 2.1). But wind pollination has not been found to be very efficient. It seems that pollination is confined mainly to individual spikes (Fig. 2.2). The fruit is a one-seeded green berry, sessile usually globose and sometimes elongated or oval. The seed is surrounded by a thin, soft pericap. During ripening, it becomes yellow and turns light red afterwards. The period taken from flowering to maturity is about 6 months. The entire life cycle of black pepper has been presented in Fig 2.5 [A-H].

Chen *et al.*, (2018) studied on the anther dehiscence, pollen viability and stigma receptivity on cultivars of black pepper (*Piper nigrum L.*). The results show that anthesis in the 10 black pepper cultivars occurred between 10.25 pm and 10.50 pm. In the pollen viability study, results suggest that pollen are more viable between 5 and 10 hours after anther dehisced. However, there are variations among the cultivars for the optimum viable stage.

Nguyen *et al.*, (2019) studied on flower biology of black pepper (*Piper nigrum*) to provide important understanding in flower biology which is vital to breeding and hybridization studies in Vietnam. Three varieties namely Vinh Linh, Phu Quoc and SRLK have been used for this study. The results showed that it takes about 242 days to 270 days from spike appearance to fruit ripening. The longest period is fruit development and fruit maturity. Anther dehiscence of Vinh Linh and Phu Quoc occurs at around 7:00 pm to 8:00 pm. However, SRLK is earlier at 4:00 pm to 5:00 pm. Sigma receptivity happens 1.8 days to 2.8 days after anther dehiscence. Stigma remains receptive from 4 days to 6 days and up to 10 days.

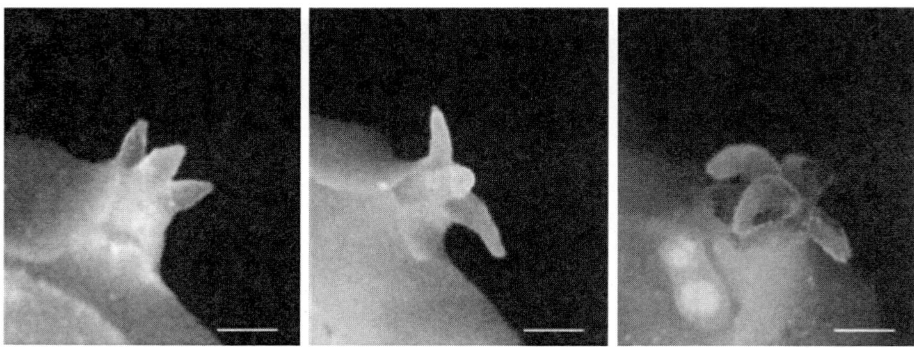

Fig. 2.1: Stages of stigma in black pepper

Fig. 2.2: Spikes in black pepper

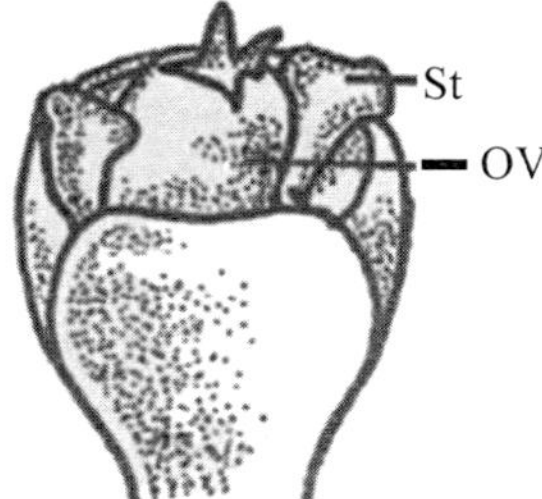

Fig. 2.3: Female flower

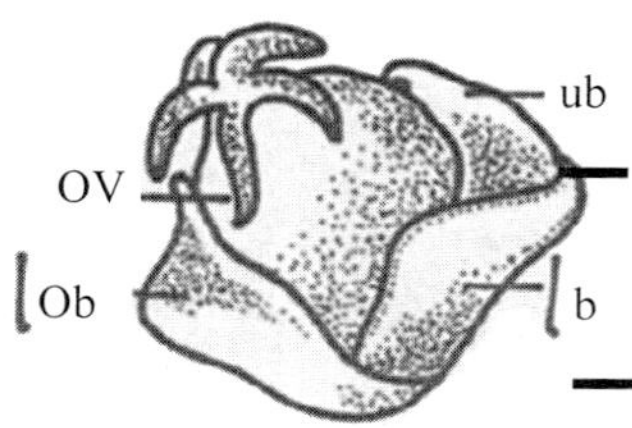

Fig. 2.4: Bisexual flower

0 day-stage1 Re-water supply; 28-30 days- stage 2 Spike appearance; 18-20 days-stage 3 Spike development; 12-13 days-stage 4 Pollen fertilization

6-9 days-stage 5 Fruit settimg; 120-130 days-stage 6 Fruit development; 34-38 days-stage 7 Fruit maturity; 25-30 days-stage 8 Fruit ripening

Fig. 2.5[A-H] Different stages of spike and fruit development in black pepper

Mode of pollination

Since pepper is mostly dioecious in wild state, cross pollination is essential for sexual reproduction in such types. Cultivated pepper, however, has mostly bisexual flowers fertilized by self pollination. Cultivated plants have been selected for genetic constancy and reliable fertility since the beginning of domestication (Frankel and Galun, 1977). Kinds of pollination, which take place in pepper are, within the same flower, between flowers in an inflorescence, and between flowers in separate inflorescence (Ibrahim *et al.*, 1985).

Varietal evaluation

Systematic research efforts in the last three decades resulted in the release of superior lines of black pepper varieties by hybridization/ open pollination/ clonal selection from the popular cultivars. Seven varieties viz., Panniyur1, 2, 3, 4, 5, 6 and 7 yielding between 1.27 and 2.57 tonnes/ha have been released by PRS (KAU) under AICRP on spices. IISR have released four varieties viz., Sreekara, Subhakara, Panchami and Pournami with an yield ranging from 2.3 to 2.8 tonnes/ha. The CPCRI, Regional station, Palode released a variety PLD-2 with a potential of 2.4 tonnes/ ha (Ravindran *et al.*, 2000).

2. Small Cardamom

Systematic position

Kingdom: Plantae

Order: Zingiberales

Family: Zingiberaceae

Genus: *Elettaria*

Species: *cardamomum*

Scientific Name: *Elettaria cardamomum* Maton.

Cytology

The chromosomal numbers for cardamom are $2n = 48$ (Gregory, 1936; Sharma and Bhattacharyya, 1939) and $2n = 52$ (Chakravarti, 1948). Cardamom is considered to be a balanced tetraploid. Allied genera such as *Globa, Balbifera, Phoemaria, Amomum* sps and *Alpinia*sps also possess $2n = 48$ and are considered to be evolved from a common basic number, $x = 12$. The *Mysore* and *Malabar* varieties of cardamom possess $2n = 50$ and $2n = 48$ chromosomes, respectively, and aneuploidy as well as structural alterations in the chromosome have contributed to the varietal differentiation (Chandrasekhar and Sampathkumar, 1986). Earlier researchers (Chandrasekhar and Sampathkumar, 1986) have reported that cardamom is of amphidiploid origin from wild species, and the two species considered to be the putative parents are the Sri Lankan cardamom *E. major* and the Malaysian species *E. longituba.*

The three natural varieties of green cardamom plants are:

Malabar (Nadan/native), as the name suggests, is the native variety of Kerala. These plants have floral racemes (which bear the pods) that grow horizontally along the ground. Mysore, as the name suggests, is a native variety of Karnataka. These plants have floral racemes which grow vertically upwards. Vazhuka is a naturally occurring hybrid between Malabar and Mysore varieties, and the panicles grow neither vertically nor horizontally, but in between. Specific characteristics of the three types have been detailed in Table 2.1.

Table 2.1: Specific characteristics of the three varieties of cardamom

Characters	var. Malabar	var. Mysore	var. Vazhukka
Adaptability	Lower altitudes 600-900 m MSL	Higher altitudes 900-1200 m MSL.	Wide range
Areas of cultivation	Karnataka	Kerala and parts of Tamil Nadu	Kerala
Plant growth	Medium	Robust	Robust
Panicles	Prostrate	Erect	Semi Erect
Capsules	Round or oblong	Bold, elongate	Round to oblong
Leaf petiole	Short	Long	Long
Capsule colour at maturity	Pale/golden/yellow	Green	Green

Source: Sastri, 1952

Cardamom *(Elettaria cardamomum* Maton) is a herbaceous perennial (2-5 m in height) having underground rhizomes with aerial pseudo stems (tillers) made of leaf sheaths. The suckers require about 10 to 12 months to attain maturity. Vegetative buds take almost 10 months to develop and for the panicle takes about an year to emerge from the newly formed tillers.

Botany

Small cardamom (*Elettaria cardamomum* L. Maton), belongs to the Zingiberaceae family. The basic chromosome number x=12 and 2n=48 indicates its balanced tetraploid nature (Madhusoodanan *et al.,* 2002).Cardamom is a tall growing, cross-pollinated, herbaceous perennial, with branched subterranean rhizomes. The real stem is the rhizome, which is subterranean in habit. Cardamom is a shallow-rooted plant. The leaves are 35 cm long and 1-10 cm wide, distichous, linear and lanceolate in shape, with short petioles. The inflorescence or the long panicles with racemose clusters arise from the underground rhizome. The flowers open in succession from the base to the top and develop into fruits. The calyx is cylindrical and persistent and the corolla tube is shortly exerted. The fruit is a trilocular capsule; the capsules are rounded. The seeds are black when fully ripe in a capsule and embedded or covered with a white mucilagenous coat. There are 15-20 seeds in each capsule with a hard seed-coat.

Floral biology

Flowers are borne on panicles, which emerge directly from the swollen base of the aerial shoot. It is a cross-pollinated plant and pollination occurs by external agents like honeybees.The panicles are erect in Cv.Mysore prostrate in Cv.

Malabar and intermediate (pendent) in Cv. Vazhukka. Panicles may be branched or simple. The peak period of panicle emergence is from November to March. Flowering normally commences from February and extends to October; May-August being the peak flowering period. After fruit set, about 90-120 days are required for the fruits to attain maturity. They are better adapted to altitudes ranging from 900 to 1200 meters from sea level and thrive well under assured, well-distributed rainfall conditions (Vijayan *et al.,* 2018). Flowers of most cardamom types and varieties are white with the central lip streaked with pink (Telja *et al.,* 2006) The labellum is oval and indistinctly 3 lobed. The calyx is cylindrical and persistent and the corolla tube is shortly exerted. Anthers are two lobed, adnate to the filament and dehisce vertically. The size of pollen grains varies from 75 to 120 μm in diameter. The stigma is funnel shaped with cilia around a small cavity. The ovary is inferior, trilocular with axial placentation and ovules are numerous in each carpel. Anthesis typically starts at 3.30 am and continues until 7.30 am, the maximum pollen bursting occurs between 5.30 am and 6.30 am. The seeds are black when fully ripe in a capsule and embedded or covered with a white mucilagenous coat.

Capsules (fruits) mature completely in about 120 days from flowering. The fruits are ellipsoidal or almost spherical, non-dehiscent, fleshy and leathery when dry. The fruit colour is green and turns golden yellow on ripening and is 1–2 cm in length. Depending upon the genotype, each capsule contains 12 to 32 seeds and the ripe seeds are black and covered with a white mucilaginous coat (Murugan *et al.,* 2016). Cardamom is highly cross-pollinated and depends on honeybees for pollination.

Pollination

Though cardamom has bisexual flowers and is self compatible, cross pollination is the rule. The flowers remain in bloom for 15-18 hours and stigma receptivity and pollen viability are maximum during 8 AM and 12 noon (Krisnamurthy *et al.*,2009). The position of stigma and anthers are such that pollination does not take place by itself without the help of some external pollinating agents. Honeybees (*Apis cerana indica* and *Apis dorsata*) visit cardamom flowers during flowering season for collecting nectar and pollen and they do help over attaining 90% of pollination.

Pollination in cardamom is entomophilous mainly by two species of honey bees viz. *Apis cerana* and *Apis dorsata.* Honey bees contribute over 90% pollination, lack of pollination causes fruit drop and fruit set in cardamom. Number of visits of bees is directly correlated with the number of seed formed in capsules. Five to six visits by bees are required for getting full seed set. At least four bee colonies (hives) / ha to are needed to enhance pollination and fruit set. Increase in yield up to nine % could be obtained by keeping bee hives.

3. Large Cardamom

Systematic position

Kingdom: Plantae

Order: Zingiberales

Family: Zingiberaceae

Genus: *Amomum*

Species: *subulatum*

Scientific Name : *Amomum subulatum* Roxb.

Large or greater cardamom is also known as Nepal cardamom and is an important spice crop of India. The seeds of large cardamom, which are commercially important. The volatile oil present in the seeds of large cardamom is one of the principle constituents responsible for its typical characteristic odour. Large cardamom (*Amomum subulatum* Roxb.) is an important perennial cash crop sympatric to sub-tropical and temperate agro-climatic zones (400–2400 m) of the Eastern Himalayan region including Nepal and Bhutan (Sharma *et al.,* 2016).

Botany

Large cardamom is a herbaceous perennial with subterranean rhizomes, which gives rise to leafy shoots and spikes. The mature plant height ranges from 1.5 to 3.0 m. Leafy shoots are formed by long sheath-like stalks encircling one another. The leaves are green or dark green, glabrous on both the surfaces, with an acuminate apex. The inflorescence is a dense spike on a short peduncle bearing 40 to 50 flower buds in an acropetal sequence. The fruit is a trilocular, many- seeded capsule. The capsule wall is echinated and is reddish-brown to dark pink in colour. Wild species of Amomum also exist in Sikkim.

Cytology

Cytology of Ammomum indicates that the diploid chromosome number of *A. subulatum* is 48. However, variability is also reported with 2n= 26, 34, 42 and 44 (Sharma and Bhattacharya, 1959).

Flowering

Large cardamom is essentially a cross pollinated crop due to the heterocyclic nature of its flowers, though they are self-fertile. Each spike bears 40-50 flowers, which open in an acropetal sequence, but only 10-15 capsules are formed per spike. The flowers remain viable for 14hr after opening. They are borne on shortly peduncled spikes of about 5-6 cm in diameter. The number of inflorescence produced on each clump ranges from 20-45, depending on the age of the clump. Each inflorescence produces 30-50 flowers. The flowers are yellowish and measure 7.03 cm in length. The most conspicuous part of the flower is the yellowish labellum/lip, which provides a platform for visiting insects. The basal parts of the petal and the labellum are fused to form a corolla tube/ nectar tube (3.07 cm long). The terminally expanded part of the labellum is 3.52cm long and 1.4 cm wide, the midrib region of the labellum is a deeper yellow and the veins are translucent. The anther is solitary, borne on a filament about 1 cm long, originating from the tip of the corolla tube, and measures 10.6 mm in length. The stamen extends beyond the anther in the form of a rolled-up leafy hood/crest. The pistil is solitary, the ovary is 5.25 mm long and contains an average of 106.8 ovules. The style is long and delicate, it passes through the groove present between the two pollen sacs. The stigma is a cup-shaped and slightly flattened, with a row of unicellular, non-receptive hairs on its margin. Only the inner surface of the cup is receptive. The stigma cup is pointed distally and the inner wall is lined with a viscous exudates. The stigma extends 1.5-2.0 mm beyond the level of the anther and is covered with the rolled-up extension of the stamen(crest) in the form of a hood. Two yellowish nectarines are located at the base of the style and fill the entire space in the lower part of the corolla tube.

Kishore *et al.* (2012) studied on floral phenology, floral visitors, foraging nature of floral visitors, nectar production, pollination efficiency and stigma receptivity of large cadamom (*Amomum subulatum* Roxb.) of the family Zingiberaceae. They opined that anthesis in large cardamom took place in the early morning (around 5:30 am) and flowers offer plenty of anthers and moderate amount of nectar to floral visitors. There were four floral visitors; bumble bee (*Apis braviceps* Smith), honey bee (*Apis cerana indica*), fruit fly (*Bactocera* sp.) and moth (*Udaspes folus*), and among them *A.braviceps* was found to be the major pollinator due to its high pollination efficiency attributed to its big body size and foraging habit, while *A. cerena* acts as a pollen robber. Each flower receives 45.92 visits of bumble bee with the foraging time of 118.93 seconds in a day. Pollination efficiency of bumble bee was as high as 100% while honey bee could pollinate only 8.41% of flower. Stigma became receptive 12 hr before anthesis (protogynous) and remains receptive even after 15 hr of flower senescence. The best time of pollination was between 6 and 9 hr on the day of anthesis.

Sharma *et al.* (2019) The average sized spikes bearing inflorescence had 45–80 flowers (62.23). The number of flowers opening each day on an average sized bush was 3.8 (range 3–6) at Lingee-Payong, 5.9 (range 4–10) at Hee-Martam, and 4.6 (range 3–9) at Jaubari. The longevity of large cardamom flowers was one day. An average number of flowers per spike per day were 3.82, 5.94, and

4.57 for Lingee-payong, Hee-Martam and Jaubari, respectively. The anthesis of the flowers took place during 5:00–6:00 h and wilted by 17:00–18:00 h. The nectar secretion starts gradually and peaks at 9:00–12:00 h, highest number of flowers was counted during this time-slot.

Main pollinators

Large cardamom is essentially a cross-pollinated crop and requires the cross-pollination of its flowers for crop production. Different species of bees are its main pollinators (Fig 2.6). However, despite its economic importance, there are very few studies on the pollination eco biology of this crop. Preliminary research by Verma (1987) confirmed honeybee, *Apis dorsata*, etc to be one of the pollinators of large cardamom by directly relating the increase in cardamom production to an increase in the number of *Apis dorsata* colonies. However, in recent years, interest in cardamom pollination is growing among the scientific community, resulting in an increase in information about its pollination requirements, pollinators, and the impact of pollination on cardamom production. A few studies have reported that common native species of bees including *Apis cerana, A. dorsata, A. florea,* and *A. laboriosa*, as well as the less noticed stingless bee, *Trigona* sp. (putka), visit large cardamom flowers (Singh *et al.*, 2012). Further studies by Sinu *et al.*, (2011) reported that *Bombus haemorrhoidalis*, *Apis cerana, Megachile lanata, Episyrphus balteatus* (hover fly), *Macroglossum stellatarum* (hawk moth), and *Aethopyga siparaja* (crimson sunbird) also visit the flowers of large

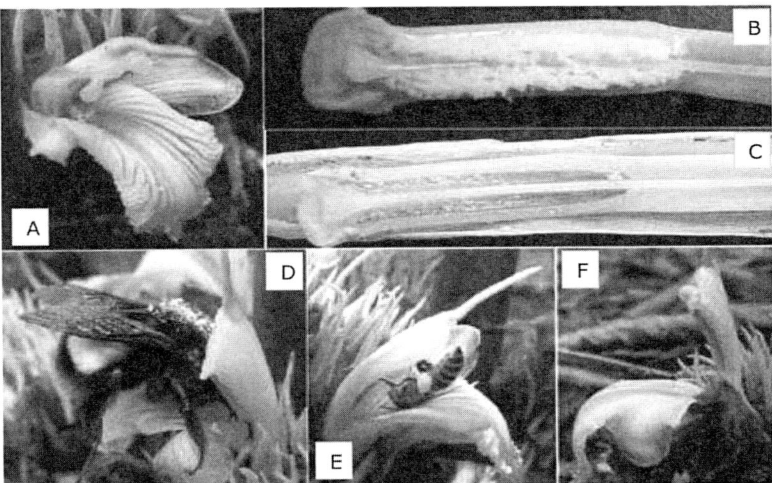

Fig. 2.6[A-F]: A, An individual flower of Amomum subulatum. **B,** Anther–stigma column **C,** Same as B, but after repeated visits by the honey bee. There is hardly any pollen left on the anther. **D,** The bumble-bee, *Bombus haemorrhoidalis* **E,** The honey bee, Apis cerana collecting pollen from the flower. **F,** A flower to show significant increase in the distance between the anther–stigma column and labellum.

4. Ginger

Systematic position

Kingdom: Plantae

Subkingdom: Tracheobionta

Superdivision: Spermatophyta

Division: Magnoliophyta

Class: Liliopsida

Subclass : Zingiberidae

Order: Zingiberales

Family: Zingiberaceae

Genus: Zingiber

Species: officinale

Botany of Ginger

Ginger has a characteristic thickened, branched rhizome which contains a brown corky outer layer and a pale yellow centre with a spicy lemon-like scent. Shoots (pseudostems) are up to 1.2 m tall arising annually from buds on the rhizome. These pseudostems are formed from a series of leaf sheaths wrapped firmly around one another with the long (up to 7 cm), narrow (up to 1.9 cm wide), mid-green leaf blades arranged alternately.

The flowering heads, borne on separate shorter stems, are cone-shaped spikes and composed of a string of greenish to yellowish leaf-like bracts. one complete flower emerges from each bract. The flowers are pale yellow in colour with a purplish edge having yellowish dots and striations. Flowering stems are rarely produced in cultivated plants. Nectar is present in the slender flower.

Floral biology

The flowers open between 14.30 and 16.30 hours and anthesis take place simultaneously. It takes about 20-25 days from the flower bud initiation to full boom. In an inflorescence blooming takes place in 23-28 days in an acropetal succession. Anthesis takes place between 1.30 P.M to 3.30 P.M. Anther dehiscence almost coincided with the flower opening or followed it immediately. Das *et al.* (1999) observed that anthesis under greenhouse and field conditions took place around 13.00-14.00 h and 09.00-10.00 h, respectively. Flowers were hermaphrodite with pin and thrum type incompatibility, and dehisced pollen grains did not reach the stigma head. No seed set was obtained by selfingor crossing.

5. Turmeric

Systematic position

Kingdom: Plantae

Subkingdom: Viridiplantae

Division: Tracheophyta

Class: Magnoliopsida

Order: Zingiberales

Family: Zingiberaceae

Genus: *Curcuma*

Species: *longa*

Botany of Turmeric

Turmeric is an upright, perennial herb with about 1 m height. The rhizome is thick and ringed with the bases of old leaves. The leaves are large, oblong, up to 1 m long, dark green on upper surface, pale green beneath. Each pseudostem bears 8-12 leaves. Flowers are yellow-white, borne on a spike-like stalk 10-15 cm long. They are sterile and do not produce viable seed. Seeds are small, ovoid, brown, not viable.

Floral Biology

Flowering in turmeric is reported to vary depending on the cultivars and climatic conditions. In *C. aromatica,* the flowering period is July-September, whereas in *C. longa,* it is September-December. Turmeric inflorescence takes 7 to 11 days to blossoming after the emergence of the inflorescence. The duration of flower opening within an inflorescence lasts for 7-11 days. Opening of the flowers took place in the morning hours around 6 AM. The anthesis starts from 7 AM and continues up to 9 AM, maximum occurring around 8 AM. Anther dehiscence takes place between 7.15 and 7.45 AM.

6. Garlic

Systematic position

Kingdom: Plantae

Subkingdom: Tracheobionta

Superdivision: Spermatophyta

Division: Magnoliophyta

Class: Liliopsida

Subclass: Liliidae

Order: Liliales

Family: Liliaceae

Genus: *Allium*

Species: *sativum*

Garlic is a bulbous perennial herb, closely related to the onion. It has a tall, erect flowering stem that reaches 2-3 feet in height. The plant has pink or purple flowers that bloom in mid to late summer. The part used medicinally is the bulb. European standards specify that garlic supplements contain not less than 0.45% allicin (Tyagi *et al.*, 2013).

Botany of Garlic

Allium sativum is a perennial flowering plant having a tall, erect flowering stem of about 1 m height. The leaf blade is flat, linear, solid, and approximately 1.25-2.5 cm wide, with an acute apex. The plant sometimes produces pink to purple flowers from July to September in the Northern Hemisphere. The bulb is odoriferous and contains outer layers of thin sheathing leaves surrounding an inner sheath that encloses the clove. Often the bulb contains 10 to 20 cloves. The garlic inflorescence is an umbel-like flower arrangement, the flower clusters arise from a common meristem. Mostly garlic produces hermaphrodite flowers.

Floral Biology of Garlic

Garlic genotypes are categorized as non-bolting, semi-bolting, and bolting. In bolting accessions, specific combinations of temperature and photoperiod significantly influence the reproductive processes (Mathew *et al.*, 2010). Long-photoperiod conditions trigger the initial elongation of flower stalks. It is pollinated by bees, butterflies, moths, and other insects.

7. Chilli

Systematic position

Kingdom: Plantae

Subkingdom: Tracheobionta

Superdivision: Spermatophyta

Division: Magnoliophyta

Class: Magnoliopsida

Subclass: Asteridae

Order: Solanales

Family: Solanaceae

Genus: Capsicum

Capsicum belongs to the family Solanaceae. It is herbaceous or semi-woody annual or perennial.

Five species are recognized mainly in chilli-

- *Capsicum frutescens*
- *Capsicum annuum*
- *Capsicum baccatum*
- *Capsicum pubescence*
- *Capsicum chinense*

Chromosome number 2n = 2x =24

C) Structure

The leaves are ovate, tapering to a sharp point, entire upto 15 cm length. The flowers are small, white and borne singly or in clusters of 2 or 3 in the axil of the leaves. The stamens are five, which are alternating the petals (Nambiar, 1987). The fruits are of diver shapes and sizes depending upon the variety.

D) Pollination

The extent of natural crossing in peppers is reported to be about 16.5 percent (Muthukrishnan *et al.*, 1993). Flowers remain open for 2-3 days and the percentage of fruit setting is 40-50 percent (Purseglove, 1977). The flowering begins 1-2 months after planting and it takes again one month for fruiting.

The flowers of chillies open in the morning between 6-10 AM and the anthers dehisce an hour after the flower opening. However, weather conditions influences much to flower opening and anther dehiscence. During cold as well as cloudy days, the opening of the flower is delayed. Crosses can be made at any time during the day, but morning hours are chosen. Flowers are emasculated in bud stage and pollen are transferred to the stigma either from mature undehisced anther by scooping it out through the side sutures with the help of needle or by touching a newly dehisced another to the stigma with the forcep. Proper washing of hands and tools (forcep, needle, scissor, etc.) are done with 95 per cent ethyl alcohol. Pollinated flowers are protected from bees by a double layer of cheese cloth, loosely wrapped around the branch, enclosing leaves and flowers, and fixed firmly. Appropriately marked plastic labels, describing the cross (i.e. female x male parent) and date are attached. Pollinated flowers are periodically checked and the cheese cloth removed after 4-6 days of pollination (Greenleaf, 1986). Fruits usually mature in about 45 days.

8. Clove

Scientific Name: *Syzygium aromaticum* (Merrill et Perry.) Syn. Eugenia carryophyllus (Spregal)

Family: Myrtaceae

Origin: Indonesia

Chromosome number: 2n=2x=22

Systematic position

Kingdom: plantae

Order: myrtales

Family: myrtaceae

Genus: *Syzygium*

Speices: *aromaticum*

Botany

The clove is an evergreen tree growing to a height of 7-15 m. The trunk is conical when young, later becoming roughly cylindrical. It begins to fork near the base, into two or three main erect branches. The smaller branches are semi-erect, the twigs are brittle, smooth and greyish-white in colour, and the whole head is bushy and dense. The leaves are simple, opposite, exstipulate, glabrous and aromatic owing to plenty of oil glands on the lower surface. The lamina is lanceolate or narrowly elliptic, bluntly acuminate at the apex, cuneate and bare. The new leaves appear in flushes and are bright pink. Later, the upper surface becomes glossy and dark green and wavy with recurved leaf margins. The hermaphrodite flowers are borne on a terminal, corymbose, trichotomous panicle. The inflorescence is shortly pedunculate, branched from the base and shorter than the leaves. The bracts and bracteoles are narrow, acute and fall quickly. The number of flowers varies from 3 to 50; a fleshy hypanthium is present, surrounded by the sepals. The hypanthium is green in the young bud, flushed pink at anthesis and turns deep red after the stamens fall. Above the sepals, there are four dome-shaped whitish petals. After fertilisation, the stamens and styles invariably fall. The lower part of the flower along with the calyx develops into a fleshy, dark, one-seeded drupe. The sepals are reduced to triangular projections and this is popularly known as the 'mother of clove'. The stamens are numerous, the anthers are pale yellow, with a small, pale brown, inconspicuous connective gland. The style is very stout, swollen at the base, pale green and dotted with glands. The two-celled, multi-ovate, inferior ovary is embedded at the top of the hypanthium. The fruit (mother-of-clove) is usually a single-seeded drupe, but occasionally contains two seeds also.

Floral biology

Flowers are hermaphrodite, borne at the terminals in small bunches. Each peduncle carries 3 or 4 stalked flowers at the end and each flower has cylindrical thick ovary clove consisting of four fleshy sepals. Above the sepals there are four whitish structures, petals, dome shaped in appearance. After fertilization stamens and styles invariably fall. The lower part of the flower, along with the calyx, develops into a fleshy, dark one-seeded drupe. The sepals are reduced to triangular projections and this is popularly known as the mother clove.

In clove anthesis takes place in the afternoon at 1:30 p.m. with a peak between 3:30 p.m. and 4:30 p.m. anthers dehisce longitudinally. Stigma receptivity was higher on the fifth day of anthesis. The best period for pollination is between fourth and sixth day after opening. Under artificial pollination the maximum fruit set obtained was 30%, while under bagged condition it was 28%. Self pollination appeared to be more probable in clove. A fertilized flower takes about three months for maturity (Thangaselvabai *et al.,* 2010).

9. Cinnamon

Systematic position

Kingdom: Plantae

Order: Laurales

Family: Lauraceae

Genus: *Cinnamomum*

Speices: *verum*

Botany

Cinnamon is a bushy, evergreen tree belonging to the family Lauraceae. It may reach a height of about 6-15 m, but in cultivation it is generally coppiced or cut back periodically. Its bark and leaves are strongly aromatic. The leaves are stiff, evergreen, with the petiole 1-2 cm long, grooved on the upper surface; the lamina is ovate or elliptic, 5-17 x 3-10 cm in area. It is strongly 3-veined from near the base, the lateral veins are 3/4 or more in length, it is reddish when young, and turns dark green above with paler veins and pale glaucous green beneath. The flowers are axillary borne in terminal panicles at the ends of twigs, the peduncle is creamy, white and 5-7 cm long. The individual flowers are very small, about 3 mm in diameter, with a foetid smell, each subtended by a small ovate hairy bract. It has 6, 9 sepals, companulate, pubescent, the stamens occur in 3 whorls with glands at the base, with hairy filaments and 4-celled anthers. The fruit is a fleshy berry, dark purple, 1-seeded, ovoid, 1.5-2 cm long with an enlarged calyx at the base. In Karnataka, the tree flowers in January and the fruits ripen 6 months later. The somatic chromosome number of *C. verum* is 2n=24.

Pollination

According to Kumari *et al.,* (2008), there are two types of plants named Type A, in which stigma becomes receptive during morning, and Type B with receptive stigma in the afternoon. The flower opening cycle is completed within a span of 12 hours as anthers become dehiscent in the evening in “Type A” and in the morning on the next day in “Type B”. Overlapping periods in “Type A” and “Type B” flowers result in cross pollination. Kumari *et al.,* (2008) determined that the majority of cultivars tested belong to “Type B’.

As reported by Purseglove (1969), flowers of Cinnamomum verum were found in axillary and terminal panicles at the end of twigs. Its peduncle is creamy white and 5–7 cm long. Individual flowers are very small, of about 3 mm in diameter, with a foetid smell. A small, ovate and hairy bract subtends from the flower. There are six campanulate and pubescent sepals. There are nine stamens in three whorls with glands at the base. The filaments of stamen are hairy and anthers are four-celled opening through four small valves. There are three staminodes. The ovary is superior and one -celled. The main reports on cinnamon floral morphology by Purseglove (1969) and Dassanayake *et al.* (1995) indicated differences.

Flowering in cinnamon is reported to have started by November and continues until the early part of March (Joseph 1981; Kubitzki and Kurz 1984; Mohanakumar *et al.* 1985). The fruits are ripen from May to June. However, according to our observation in southern Sri Lanka, off season flowering occurs in March and April, and occasionally in August and September.

10. Allspice

Systematic Position

Kingdom: Plantae

Phylum: Spermatophyta

Subphylum: Angiospermae

Class: Dicotyledonae

Order: Myrtales

Family: Myrtaceae

Genus: Pimenta

Species: dioica

Botany

Allspice is a small, functionally dioecious evergreen tree, 7-10 m tall, slender trunk profuselybranched at their extremities. Bark, smooth and shiny, pale silvery brown, shedding in strips of 25-75 cm long at intervals. Leaves, borne in clusters at the ends of the branches, simple, opposite, aromatic when crushed. Flowers structurally hermaphrodite, but functionally male or female,white, aromatic, 8-10 mm diameter.

11. Nutmeg

Cytology: 2n=42

Systematic position

Kingdom: Plantae

Phyllum: Tracheophyta

Class: Magnoliopsida

Order: Magnoliales

Family: Myristicaceae

Genus: *Myristica*

Species: *fragrans*

M. beddomi

M. malabarica

M.fragrans

Fig. 2.7: Different species of Myristica

Morphology

- Evergreen tree with dense foliage
- Trees are 10 to 12 meter high and branches are spreading with a dark grey bark.
- It is a doiecious plant. Male trees have erect branches and the leaves are generally smaller in size and conspicuously less leafy than female trees.
- The calcium oxalate content of the leaves is also taken as a criteria for identifying the sex in nutmeg.
- Leaves are shiny and oblong to oval in shape.
- Flowers appear in cymes.
- Flowers are small, pale yellow and bell shaped slightly aromatic.
- Fruits are fleshy, globose in shape and lemon yellow to light brown in colour.
- Fruits ripen 9 months after flowering. Inside the fruit there is a single, glossy brown seed with a brittle shell over which there is a beautiful, brilliant, scarlet, netlike, (reticulated) membrane or aril known as the mace, which is very fragile and aromatic.

Sex and Spacing

Fig. 2.8: Flower and different parts

Commercial nutmeg (*Myristica fragrans*) seedlings germinate in a 1:1 sex ratio, the result of purported sex chromosomes [22], but whether this is true for all nutmeg species remains unknown. In nature nothing close to a unitary sex ratio has been found, but this is not unexpected, considering that dioecy may allow for intersexual differences in resource use, longevity and survival, or aggregation for increasing pollination efficiency. Female-biased ratios ranging from 10:1 to 20:1 in plantations do not seem to negatively affect fruit production, so even in such female-skewed populations pollen is not limiting, although lack of pollinators, both in number and species diversity, may affect fruit set .

However, the spacing of trees in a plantation is much closer than in nature. In Australia, the mean distance from male to nearest female for *Myristica insipida* was 9.6 m and 19.4 m in two different forest communities, but these distances may represent clustering in mesic habitats. In Amazonia, the average distance to the nearest flowering conspecific ranged from 57 to 147 m in four nutmeg species. No one knows how nearest neighbor distances affect pollination, but it certainly depends upon the pollen vector (Sharma *et al.*, 2013).

Flowering

- Flowering pattern of male and female trees is different. In female trees, flowering continued to seven months, whereas in male trees, flowering was observed throughout the year.
- Highest flowering in both Male and female trees is maximum in July followed by October,which will be more favourable time for better pollination and fruit set.

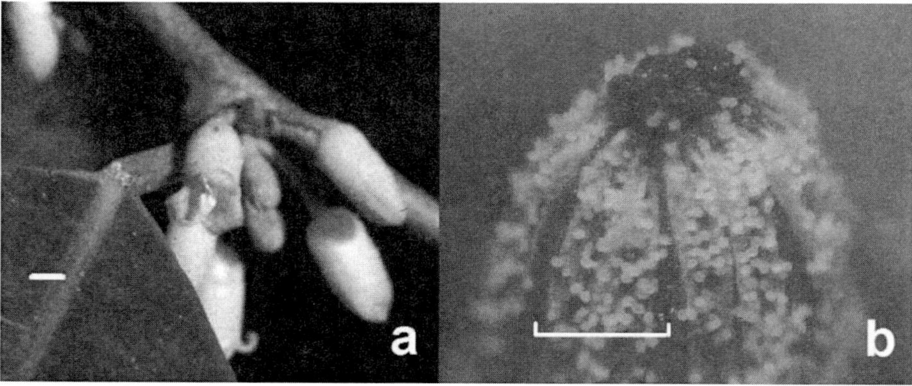

Fig. 2.10: Male flowers and fully opened anther

Anthesis

- The anthesis of male flowers is maximum between 7pm to 1am. Anther dehiscence occurred about 24 hours prior to anthesis.
- The anthesis of female flowers is peak between 9pm to 3am. The flowers are receptive from the day of anthesis up to 6 days.

Pollination

- The mode of pollination is cross pollination.
- The pollination in case of Nutmeg takes place through wind and insects(thrips).
- The first scientific study of nutmeg flowering and pollination was carried out in commercial nutmeg plantations in lowland Kerala, India. Both beetle and thrip floral visitors were observed, but only one species of small beetle (Anthicidae) was a good candidate for being an effective pollinator. This study provided little useful information about the pollinators of commercial nutmeg elsewhere, and since the study was not conducted in a natural forest community, the biology and pollinator interactions of wild nutmegs, the full diversity of pollinators, and the organization of natural nutmeg populations, remained unknown.
- In Myristicaceae, thrips were reported as floral visitors of Neotropical species, and thrips, beetles, and flies have been found effecting pollination.

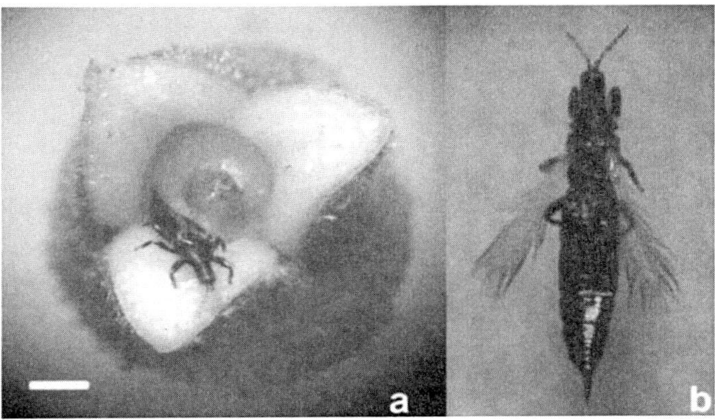

Fig. 2.11: Floral vectors

Fruit set

- Highest fruit set is in trees on western and eastern aspects(Nazeem 1979). The fruits attained maturity in 206 to 237 days after fruit set. The developing fruits followed a sigmoid growth pattern (Selvakumar *et al.*, 2011).

3

Centres of Origin

Since historical times, Indian spices have been contributed not only to the country's wealth and prosperity but also to the subcontinent itself as a whole. Spices precisely witnessed many wars, bloodshed, armoury alongwith the glowing accounts and adventurous voyage by Pliny, Ptolemy, Marco Polo, Vasco da Gamma and other travelers. The Southern parts of India, especially the Malabar Coast, with its abundance of the World's best pepper, cinnamon and ginger was enormously creating impact of a plethora of spices which attracted the foreign trade and continue even today by creating a unique brand value globally. In the past, invaders travelled to India by overcoming many odds and barriers with the prime focus on capturing spice trade and dominating the trade scenario of those days throughout the world. As a result, gradually Indian spices travelled far and wide to traditional destinations by providing flavor, aroma, pungency, taste and colour to a variety of foods and beverages.

It was the end of 18th century, from when scientific research began on the idea of origin of cultivated plants. The pioneers of this field were Darwin, Alphonse de Candolle, N.I Vavilov and many more. Charles Darwin wrote his famous book 'On the Origin of Species by Means of Natural Selection', or the 'Preservation of Favoured Races in the Struggle for Life' published on 24th November 1859. According to him, cultivated plants may have originated from its wild form after a series of intense modifications. Afterwards, a Swiss botanist Alphonse de Candolle studied 247 cultivated plant species and documented his work in the book "Origin of cultivated plants" in 1886. Later, Vavilov first identified the Centres of origin. He could determine the origin of plants by analyzing the pattern of variation and proposed that, crop plants have been evolved from their wild species in the geographic region of the greatest genetic diversity of that particular plant. That area is known as the primary centres of origin of the particular plant. But, some crop species were found to show substantial genetic diversity even though they did not originate there. Such areas are known as secondary centres of origin for that particular species.

For crop plants, Nikolai Vavilov identified varying numbers of centres : three in 1924, five in 1926, six in 1929, seven in 1931, eight in 1935 and seven again in 1940 (Vavilov and Love,1992; Corinto, 2014). Vavilov argued that plants were not domesticated somewhere in the world at random, but that there were regions where domestication started. The centre of origin is also considered as the centre of diversity.

Ultimately eight main centres with some subcenters were proposed by Vavilov. These centres were:

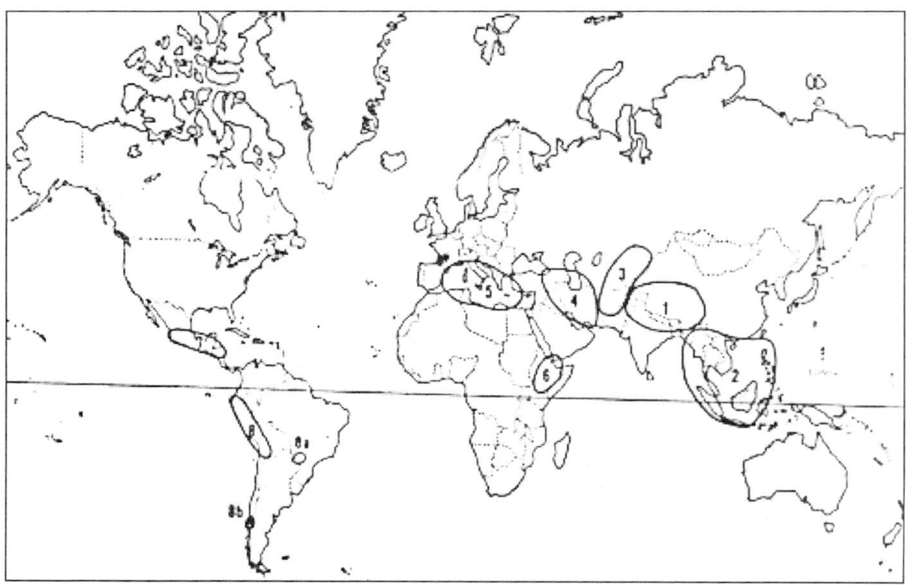

Fig. 3.1: Centres of origin

1. **The China Centre**: It consists of the mountainous regions of central and western China and the neighbouring low lands. It is the largest and oldest independent centre.
2. **The Hindustan Centre**: This includes Burma, Assam, Malaya, Java Borneo, Sumatra and Philippines, but excludes North West India, Punjab and North West Frontier Provinces.
3. **The Central Asia Centre**: It includes North West India, all of Afghanistan, Tadjikistan, Uzbekistan and western Tian-Shan. It is also known as *the Afghanistan Centre of Origin.*
4. **The Asia Minor or Middle East Centre**: This is also known as the *Near East or the Persian Centre of Origin*. It includes the interior of Asia Minor, the whole of Transcaucasia, Iran and Highlands of Turkmenistan.

5. **The Mediterranean Centre**: This includes the borders of the Mediterranean Sea.
6. **The Abyssinian Centre**: It includes Ethiopia, Eritrea and part of Somalia.
7. **South Mexican and Central American Centre**: This includes South Mexico, Guatemala, Honduras, Costa Rica and Central America.
8. **The South American Centre**: This centre includes the high mountainous regions of Peru, Bolivia, Ecuador, Colombia, Sothern Chile, Brazil and whole of Peraguay.

Table 3.1: Origin of some important spice crop(s)

Sl No	Spice crop (s)	Botanical name(s)	Family	Origin
1	Allspice	*Pimenta diocia*	Myrtaceae	West Indies, Central and South America
2	Aniseed	*Pimpinella anisum*	Apiaceae	Mediterranean region
3	Cardamom	*Elletaria cardamomum*	Zingiberaceae	Western ghats of India
4	Celery	*Apium graveolens*	Apiaceae	Southern Europe
5	Cinnamon	*Cinnamomum verum*	Lauraceae	Ceylon and Malabar coast
6	Clove	*Syzygium aromaticum*	Myrtaceae	Moluccas
7	Coriander	*Coriandrum sativum*	Apiaceae	Mediterranean region
8	Cumin	*Cuminum cyminum*	Apiaceae	Upper Egypt, Turkey and Eastern Mediterranean region
9	Fenugreek	*Trigonella foenum-graecum*	Apiaceae	Egypt
10	Fennel	*Foeniculum vulgare*	Apiaceae	Southern Europe and Mediterranean region
11	Garlic	*Allium sativum*	Amaryllidaceae	Central Asia and Mediterranean region
12	Ginger	*Zingiber officinale*	Zingiberaceae	South East Asia
13	Nutmeg	*Myristica fragrans*	Myristicaceae	Moluccas, East Indies
14	Pepper	*Piper nigrum*	Piperaceae	Western Ghats, Upper Assam Forests
15	Saffron	*Crocus sativus*	Iridaceae	Greece and Asia Minor
16	Turmeric	*Curcuma longa*	Zingiberaceae	South East Asia
17	Tamarind	*Tamarindus indica*	Fabaceae	Eastern Tropical Africa
18	Vanilla	*Vanilla planifolia*	Orchidaceae	Mexico to Brazil

4

Modes of Reproduction

Cultivated plants in general including the spice crops observed today have been evolved through selections from their weedy and wild types. Apart from this, there had been another force which could bring the wild forms into cultivated forms named human intervention which is actually the process of domestication. Natural mutations and genetic recombination after selfing and crossing also played a significant role in the entire process. Whereas, natural forces take a long time to come into act, the techniques of plant breeding was devised and fine tuned throughout passing of times, essentially to accelerate the speed of evolution using these natural phenomenons. It was, therefore, mandatory for us to know the modes of reproduction and pollination to create variability in natural population and make all thise happen to what we see today in organised cultivated forms. Moreover, knowledge on reproduction of different plant species helps the breeder to choose the right breeding procedure to be used for crop improvement. Modes of reproduction help to evaluate the genetic constitution of crop plants by determining the extent of homozygosity or heterozygosity which, in turn, are unanimously applicable to any plant. To comprehend modes of reproduction in spice crops effectively, understanding about the floral biology is absolutely necessary. Plant reproduction, a process by which one species gives birth to its offsprings, may be of two types:

1. Sexual reproduction
2. Asexual reproduction

4.1 Sexual Reproduction

When the offsprings are developed via zygotic embryos produced through the fusion of male and female garments (fertilization), it is known as sexual reproduction. Plants which are propagated through seeds belong to this category. Development of new organism via zygotic embryo in plants is precisely dependent on biological processes known as sporogenesis, gametogenesis and fertilization.

Sporogenesis

It is the production of sporogeneous cells, i.e., microspores and megaspores. When sporogenesis is related to the production of male gametes, it is known as microsporogenesis and when the process is related with the production of female gametes, it is known as megasporogenesis.

Microsporogenesis

It is the process of alteration of generation where meiotic division of the sporophytic cells in the pollen sacs of anther occurs to form haploid microspore (known as microspore mother cells, MMC or pollen mother cell, PMC). Each pollen mother cell produces four microspores and each microspore ultimately gets transformed into pollen grains after thickening of the wall.

Megasporogenesis

It is the process of alteration of generation where meiotic division of a single sporophytic cell within the ovule occurs to form haploid megaspore (known as megaspore mother cell, MMC). Each megaspore mother cell ultimately results in a single functional megaspore after degeneration of three sister megaspores.

Gametogenesis

The production of male and female gametes in the microspores and megaspores is known as gametogenesis.

Microgametogenesis

It is the process of development of pollen grain from the microgametophyte in the microspore mother cell within the anther of the plant. On maturation of the pollen, the microspore nucleus divides mitotically to produce a generative and a vegetative or tube nucleus. during pollination, the pollen reaches the stigma and starts to germinate. The pollen tube, on entering the stigma, moves down the style. The generative nucleus at this phase undergoes another mitotic division to produce two male gametes or sperm nuclei. The pollen along with the pollen tube possessing a pair of sperm nuclei is called microgametophyte. The pollen tube enters the embryo sac through micropyle and discharges the two sperm nuclei.

Megagametogenesis

It is the process of development of embryosac which is the female gametophyte from the megaspore. The nucleus of the functional megaspore undergoes three mitotic divisions toproduce eight nuclei. Three of these nuclei move to one pole and produce a central egg cell and two synergid cells on either side. Another three nuclei migrate to the opposite pole to develop into three antipodal cells. The megaspore thus develops into a mature female gametophyte called megagametophyte or embryo sac (the process is known as megagametogeneis).

Fertilization

It is the fusion of haploid male and female gametes to form diploid zygote. Mitotic division of zygote forms the embryo.

4.2 Asexual Reproduction

Propagation or multiplication of plants without union of male and female gametes is known as asexual reproduction. Asexual reproduction may occur both by propagation through vegetative plant parts or fromapomictic embryos developed without sexual fusion. There are some specialized plant parts identified to promote vegetative reproduction viz., underground stems, sub aerial stems, roots and bulbils. Here are some examples of some vegetatively propagated spice crops.

Rhizome: Turmeric (*Curcuma longa*), Ginger (*Zingiber officinale*)

Bulb: Garlic (*Allium sativum*), Onion (*A. cepa*)

Sub aerial stems (runner, sucker, stolon, etc.): Mint (*Mentha sp*).

Bulbils (modified forms of flower which when fall on the ground, develop into plants): Garlic (*Allium sativum*)

Apomixis

It is a special phenomenon of asexual reproduction where seed is obtained as the ultimate output. Here embryo is formed from the egg or other cells of the embryo sac or its surroundings without sexual fertilization (fusion of gamets). Thereby, apomixis is known to produce maternal clones. There are four types of apomixis: viz.

Parthenogenesis: development of embryo from the egg cell without fertilization.

Apogamy: development of embryo from either synergids or antipodal cells of the embryosacwithout fertilization.

Apospory: In this case, embryosac is developed without reduction from diploid cell of ovule lying outside the embryosac. Then embryo is developed directly from the diploid egg cell without fertilization.

Adventive embryony: development of embryo directly from the diploid cells belonging to either nucellus or integuments is referred to as adventive embryony.

5

Pollination and Emasculation Techniques

Pollination

Pollination is the act of transfer of pollen from the male reproductive organ of a plant to the female reproductive organ of same or another plant.

Pollination is of two types

5.1 Self Pollination

It is the transfer of pollen from the anther to the stigma of the same flower from where the anthers have been generated. Thus, self pollination happens to be denominated as autogamy where auto means "same" and gamy means "union". Obviously, self pollination is found in bisexual flowers. But self pollination also occurs when pollen is transferred from one bisexual flower to the stigma of another bisexual flower of the same plant. This kind of autogamy is known as geitonogamy. So, self pollination has to be defined keeping the parent plant on the main focus not the flower. If genetical consequences are considered, self pollination promotes the occurance of purelines or homozygotes. Self pollinating species developed themselves evolutionary fit due to homozygous balance and do not exhibit significant inbreeding depression.

Natural mechanism to promote self pollination

a. **Bisexuality:** Occurrence of both the male (androecium) and female (gynoecium) reproductive parts in the same flower is known as bisexuality.

b. **Homogamy:** It is the state ofmaturation of anthers and stigma of a flower at the same time is called homogamy. Homogamy is essential make the self pollination effective.

c. **Cleistogamy:** Occurrence of pollination and fertilization in the unopened flower bud is known as cleistogamy. It is one of the most effective mechanisms which promote self pollination.

d. **Chasmogamy:** In chasmogamy, flowers only open after the completion of pollination. This mechanism also promotes self pollination.

5.2 Cross Pollination

It is the transfer of pollen from the anther of one flower to the stigma of another flower physically borne on different plant. Different 'plant' necessarily means plants of the same or different species. Naturally unisexual flowers are cross pollinated. That is why cross pollination happens to be denominated as allogamy where 'allo' means "different" and 'gamy' means "union". Theoretically, cross pollination should result in union of two gametes which are genetically dissimilar. So, occurrence of mating between individuals who are ancestrally related is known as sibmating. Sib mating also promotes the occurance of purelines or homozygotes. Cross pollination principally promotes the occurrence of heterozygotes. Cross pollinated species show considerable inbreeding depression upon selfing.

Natural mechanism to promote cross pollination

a. **Dicliny:** It is the occurrence of either male (androecium) or female (gynoecium) reproductive part in the same flower. This is of two types: viz. i) monoecy and ii) dioecy.

   If unisexual male and female flowers are borne separately in the same plants, it is known as monoecy. Male and female flowers may or may not share a common inflorescence of the same plant.On the contrary when unisexual male and female flowers are borne in different plants, the situation is denominated as dioecy. For a better understanding, in case of dioecy, male flower bearing plant can be designated as male plant whereas female flower bearing plant can be designated as female plant. But male or female plant concept is not valid for monoecy.

b. **Dichogamy:** It refers to the condition when maturation of anthers and stigma of the same flowers occur at different times. Dichogamy is the efficient tool to promote cross pollination when the flower is of bisexual type. Dichogamy is of two types: viz. i) protogyny and ii) protandry. When female reproductive unit matures before the male, the situation is known asprotogyny. When male reproductive unit matures before the female, the situation is known as protandry.

c. **Heterostyly:** It indicates the difference in morphological forms to promote cross poliination where styles and filaments in a flower are of different lengths.

d. **Herkogamy:** Impediment to self-pollination due to some physical hindrancelike the presence of hyline membrane covering the anther is known as herkogamy. As a result,dehiscence of pollen grains is restricted and thereby self pollination is prevented.

e. **Self incompatibility:** This mechanism doesn't support the fertilization event to happen with the self pollen (pollen coming out from the same flower or same plant or pollen with identical genotype) when that very pollen is biologically fertile.Therefore, it prevents self-pollination and promotes cross pollination.

f. **Male sterility:** In some species, sometimes individuals are found where the pollen grains are biologically non-fertile i.e. not able to fertilize any egg cell. Certainly, thoseplants have to depend on other individuals who are capable of producing fertile pollen grains.Henceforth, there is no question of self pollination. Such condition is known as male sterility which exclusively prevents self-pollination and promotes cross pollination. Its control is of three types: viz. genetic, cytoplasmic and cytoplasmic genetic. Male sterility is scientifically exploited in commercial hybrid seed production where hand emasculation and pollination is either not possible or too much tedious job to do.

Methods of pollination in spices

Pepper: Black pepper is a predominantly self pollinated. While various degree of protogyny is encountered. Positive geotropism, spatial arrangement of flowers, sequential ripening of the stigma and non-chronological dehiscence of anthers stimulate selfing. Selfing with occasional outcrossing is the predominant mode of pollination in cultivated bisexual black pepper. It is adapted for pollination by gravitational descending of pollen grains combined with action of rain water or dew drops.

Cardamom: Though cardamom has bisexual flowers and is self compatible, cross pollination is the rule. Self-pollination is hindered due to the slight protrusion of the stigma above the stamens. In cardamom, cross pollinationis mediated by activity of the bees (*Apiscerana*, *Apis indica* and *Apis dorsata*) as pollinators. The active for aging of bees is seen in the morning hours of the day which help in increasing the fruit set of cardamom.

Vanilla: *Vanilla planifolia* usually flowers onlyonce in a year over a period of about 2-3months. In India the peak flowering period isduring January to March. Irrespective of the fact that vanilla flower is self –fertile the floralarchitecture of vanilla is such that natural pollination is impossible, due to the separation of anther from stigma by the rostellum. A pollinating agent is essential for transferring pollen from anther to stigma. In Mexico andCentral America, where vanilla is indigenous some of the flowers are pollinated the bees of the genus *Melipona* and also humming birds. Elsewhere hand pollination is unavoidable for fruit set.

Tree spices: Nutmeg is a dioecious tree and the chief agent of pollination is wind. In cinnamom flowers are small; numerous in terminal and axillary panicles on current seasons growth. They are cross-pollinated insects are the pollinating agents. In clove the flowers are bisexual and self-pollination is predominant in clove.

Seed spices: In coriander the inflorescence is compound umbel and the hermaphrodite and staminate flowers are small. The flowers are protandrous. Honey bees are the main pollinating agent. In fenugreek the flower is leguminous, which is small and borne onraceme and is extremely difficult to manipulate for artificial emasculation and hybridization. It is a typical self-pollinator in which fertilization occurs within the unopened floral buds (Cleistogamy).

How to identify the mode of reproduction of a plant?

Knowledge about the mode of reproduction of any plant can be gathered by scientific observation over time for a particular location even if the species is unknown to the observer. First of all, if the species doesn't flower at all, the species is hypothesized to be vegetatively propagated. If, flowering is observed, then there may be number of conclusions according to the observations which are as follows

Observation	Conclusion about mode of reproduction
1. Flowers may either be perfect or imperfect.	There is high probability that, the flower is self pollinated.
1.1 If perfect, it is bisexual or hermaphrodite condition i.e., both the male and female reproductive units are present.	If, any of these mechanism viz., dichogamy or herkogamy or male sterility or self incompatibility is operating in that perfect flower, chance of cross pollination will be high. Confirmation about selfing or crossing can be done by bagging the perfect flower. Bagging would restrict the entry of foreign pollen. So, if no seed setting is observed in a bisexual flower after bagging, the plant is certainly cross pollinated.
1.2 If imperfect i.e., either of the male or female reproductive unit is present. Both monoecious and dioceious type bear imperfect flower.	Both monoecious and dioecious flowers favour cross pollination. But, chances of self pollination are there in monoecism if geitonogamy is happening. If there is no seed setting after bagging imperfect flower bearing plant (do not bag the flower only as in case of perfect flower), then cross pollination is the rule.

Techniques of artificial hybridization

For the creation of new variability,artificial hybridization is the most popular technique. Variability offers plant breeder the power of selection. Therefore, crossing dissimilar genotypes is a routine job of the breeder. There are two kinds of needs viz., doing selfing in cross pollinated crops to make inbreeds as well as crossing in self pollinated crops to create heterozygous outbreeds. To do so, breeder has to break the natural barriers of pollination. In case of cross pollinated crops, breeder has to restrict the entry of foreign pollen entry by bagging the flower and facilitate the pollination by applying the self pollen of pollen of interest. In contrast, for making outbreeds of self pollinated crops, breeder has to restrict the pollination by self pollen discarding the male reproductive unit hours before its anthesis of the flower and facilitate the pollination by applying the pollen of interest. Although, selfing as well as crossing techniques may vary from species to species. If cross pollination is controlled by self incompatibility or breeder is trying to make wide hybridization, sometimes, bud pollination of stylar pollination is adapted to achieve selfing. Bagging of whole plant instead of flower is done wherever there are the chances of insect pollination to avoid uncontrolled crossing.

Methods of Emasculation

Emasculation essentially means making the organism less masculine. In plant breeding, it is the practice of removing stamens or anthers or rather destroying the pollen in order to compel the flower act solely as female by adapting some artificial methods. Emasculation is done both in bisexual and monoceous flowering condition. The breeders' objective of emasculation is to perform pollination of own choice. Hence, this technique is a prerequisite for making preferred cross combination. It is equally important both for research as well as hybrid seed production purpose. In bisexual flowers, only male sex organ preferably anthers are deliberately cut out to prevent any chance of self pollination. Several techniques of emasculation are available. Depending on the flower size, appropriate technique is adapted. Hand emasculation is preferred in manageable and in a bit large flowers.

1. *Hand Emasculation*

As the anthers of a flower are manually removed or discarded preferably with the help of forcep, it is so named. Generally hand emasculation is done in relatively large flowers because flowers are easy to hold in-between fingers. But, removal of anthers must be done hours before the time of dehiscence of anthers of that species. Time of anthesis varies species to species. To obtain the best result, emasculation may be practiced in buds/blooming buds/ young flowers. Surrounding young flowers situated close to the emasculated flower should be removed and emasculated flower must be bagged to avoid uncontrolled pollination and tagged for proper identification. While doing emasculation, the gynoecium should not be injured or damaged at any cost. Hand emasculation is practiced both for research and commercial hybrid seed production purpose.

2. *Suction Method*

Hand emasculation is not applicable for many species which produce very small flowers. Hence, they are emasculated just after the opening of flowers early in the morning. Anthers are sucked out from the flowers with the help of glass or rubber tube which is attached to a suction hose. Suction pressure is very crucial here because the breeder has to save the female floral units. The main drawback of this method is the chances of self pollination which may go upto 10% or more. To reduce the probability of selfing, stigma may be washed by a flush of water.

3. *Hot Water Treatment*

The whole inflorescence is dipped in hot water before the anthers dehisce preceding the opening of the flower. This method has been devised based on the concept that, anthers are susceptible than female floral units towards external as well as genetic factors. Therefore, exposure of pollen grains to high temperature kills them while female floral units remain alive. But, exposure temperature and duration both are very important here. Hence, temperature of water and duration of treatment vary from species to species.

4. *Alcohol Treatment*

Immersing of the whole inflorescence in alcohol having standardized concentration for a brief period followed by rinsing with water is done here. This procedure of chemical emasculation is better over suction method although this is not a popular one.

5. *Cold Treatment*

Cold treatment like hot water treatment kills the pollen grains without damaging gynoecium. In the case of rice, treatment with cold water 0.6°C kills the pollen grains without affecting the gynoecium. This is less effective than hot water treatment.

6. *Genetic Emasculation*

Genetic/ cytoplasmic male sterility may be used to eliminate the process of emasculation. This is useful in the commercial production of hybrids in maize, sorghum pearlmillet, onion, cotton, and rice, etc.

In many species of self-incompatible cases, emasculation is also not necessary, because self-fertilization will not take place. Protogyny will also facilitate crossing without emasculation (e.g.) Cumbu [*Pennisetum glaucum* (L) R. Br.].

7. *Use of Gametocide*

Also known as chemical hybridizing agents (CHA) chemicals which selectively kills the male gamete without affecting the female gamete. eg. Ethrel, Sodium methyl arsenate, Zinc methyl arsenate in rice, Maleic hydrazide for cotton and wheat.

Bagging

Immediately after emasculation the flower or inflorescence is enclosed with suitable bags of appropriate size to prevent random cross-pollination.

Crossing

The pollen grains collected from a desired male parent should be transferred to the emasculated flower. This is normally done in the morning hours during anthesis. The flowers are bagged immediately after artificial crossing.

Tagging

The flowers are tagged just after bagging. They are attached to the inflorescence or to the flower with the help of a thread. The following may be recorded on the tag with pencil.

6

Classical Methods of Breeding

6.1 Introduction

Plant breeding is the indispensible component of elevating production and productivity of spices by evolving high yielding and/or with good quality type. Thereby, the practice of identification and selection of desirable traits and introgression of those characters in a single plant is being done since the beginning of organized agricultural activities by humankind. The act of plant breeding was reshaped and got a momentum with a scientific basis after the discovery and rather rediscovery of Mendel's laws. Worldwide, humankind could enjoy the direct benefits of methodical plant breeding during green revolution. Manipulation of genetic architecture for creating variation in the population and successive selection is the basic of all kinds of crop breeding. But the process of creation of variability and selection differs from crop to crop depending on their mode of reproduction. It is comparatively easier to get a new homogeneous population in self pollinating plants. Therefore, segregating generations are handled differently in these two categories of self and cross pollinated crops.

6.2 Plant Breeding Methods

Suitable breeding procedure to be adapted to specific condition is always determined from the reproductive behavior of the particular crop. Based on the mode of pollination and reproduction crop plants are divided into three groups *viz.*

1. Self pollinated
2. Cross pollinated
3. Vegetatively (asexually) propagated

Self Pollinated Species

Self-fertilizing species are known as self-pollinated or autogamous species. In these plants, development of seed takes place by self pollination (autogamy). Hence self pollinated species are also known as inbreeders. Various plant characters such as homogamy, cleistogamy, chasmogamy, bisexuality etc. promote self fertilization.

Some important features of autogamous species are

1. Regular self pollination occurs in them
2. They are homozygous *i.e.*, they are true breeding type.
3. Generally do not have recessive deleterious genes, because deleterious genes are eliminated due to inbreeding by way of gene fixation.
4. In autogamous species, new gene combinations are not possible due to self pollination.

Cross pollinated species

Cross fertilizing species produce seed by cross pollination (allogamy) hence, also known as allogamous species or out-breeders. Different flower characters that promote cross pollination include dichogamy, monoecy, dioecy, heterostyly, herkogamy, self incompatibility and male sterility.

Some important features of out breeders are

1. They are random mating type. In such population (panmictic), each genotype has equal chance of mating with all the other genotypes.
2. Individuals are heterozygous in their genetic makeup.
3. Individuals have deleterious recessive gene which are hidden by masking effect of dominant genes.
4. Out breeders exhibit high degree of inbreeding depression on selfing.
5. Cross pollination allows new gene combinations.

Asexually propagated species

Asexually propagated plants are multiplicated by means of vegetative plant parts like stem or root cuttings or by others. These species are found in both self and cross pollinated groups. Generally asexually propagated species are highly heterozygous and have broad genetic base with wide adaptability.

6.3 Breeding Methods for Self Pollinated Crops

Domestication and introduction

Domesticated food crops are derived from a phylogenetically diverse assemblage of wild ancestors through artificial selection for different traits.

Our understanding of domestication, however, is based upon a subset of well-studied 'model' crops, many of them from the Poaceae family. Many typical features of domestication associated with model crops, including changes in ploidy level, loss of shattering, multiple origins, and domestication outside the native range, are less common within this broader dataset. In addition, there are strong spatial and temporal trends. The frequencies of some domestication syndrome traits (e.g. non shattering) have decreased over time, while others (e.g. changes to secondary metabolites) have increased. The influences of the ecological, evolutionary, cultural and technological factors are enormous that make domestication a dynamic and ongoing process.

Plant domestication fundamentally altered the course of human history. The adaptation of plants to cultivation was vital to the shift from hunter–gatherer to agricultural societies, and it stimulated the rise of cities and modern civilization

The evolution of domesticated forms of plants involved the selection of traits that were suited to the human rather than the wild environment. The types of traits that are selected have been similar across different species plants giving rise to the concept of the domestication syndrome. Some controversy still persists about the nature of selection of such traits and the degree of human consciousness involved, although the majority of the field accepts that most traits were probably subject to unconscious selection. It has become apparent in recent years that understanding the nature of the plurality of processes underlying the domestication syndrome is key to understanding the origins of domestication.

6.3.1 Mass selection

Mass selection is one of the oldest breeding method used for both self and cross pollinated crops. From the very beginning of agriculture, farmers used to harvest seeds from good looking plants of their field and save them for growing crop next year. That is nothing but mass selection. The mass selection is selection of plants from a genetically variable population based on their phenotype and their seeds are bulked without progeny testing. Here the base population is a heterogeneous variable group consists of several line, some of them are good and some are bad with respect to particular trait of interest. So the main objective of mass selection is to increase the frequency of desirable genotypes in the new improved population by selecting superior plants. As no new variability is created in the original population this method exploit the existing variability of the population. So the cultivar developed through this method is improved with respect to its average performance and uniformity over the base population. Mass selected cultivars are uniform mostly for easily selectable qualitative traits whereas considerable amount of variation is present for quantitative traits which lead to greater stability against abiotic and biotic stress conditions.

Mass selection has the following applications:

Variety purification: With time an existing variety (self-pollinated) may lost its purity due to outcrossing, mutation or mechanical seed mixture. Mass selection is used to maintain purity by discarding the offtypes/variants that deviates in characteristics from original variety.

New cultivar development: New cultivar is developed from the base population by selecting similar type superior individuals that give rise to improved average performance of the population.

Based on the selection procedure of individuals, mass selection is categorized as positive mass selection and negative mass selection. In positive selection a number of desirable plants uniform for the trait(s) of interest are selected. Whereas in negative mass selection only undesirable plants are rogued out and rest are bulked for advancing the next generation.

6.3.2 Procedure of mass selection

Year 1/Season 1: Select phenotypically uniform, desirable plants from the base population. The base population may be germplasm collection, landrace that must have genetic variation. The selected plants must be uniform for different visible agronomic traits like height, flowering time, maturity, seed size, colour etc. for better farming operations and acceptabity to the farmer. The selected individuals are harvested and seeds are bulked. Adoption of mechanical device sometimes may augment the selection process. For example, if similar size and shape of seeds are to be selected then sieves of particular mesh size can sort seeds of uniform size.

Year 2/Season 2: The composited seeds are planted in replicated yield trial and evaluated using the original population as check. This test may be repeated over years at different locations for checking uniformity and adaptability over different environments. Along with this the newly developed population may be evaluated for disease and pest susceptibility and quality parameters of the economic plant part. The procedure is given in Fig 6.1.

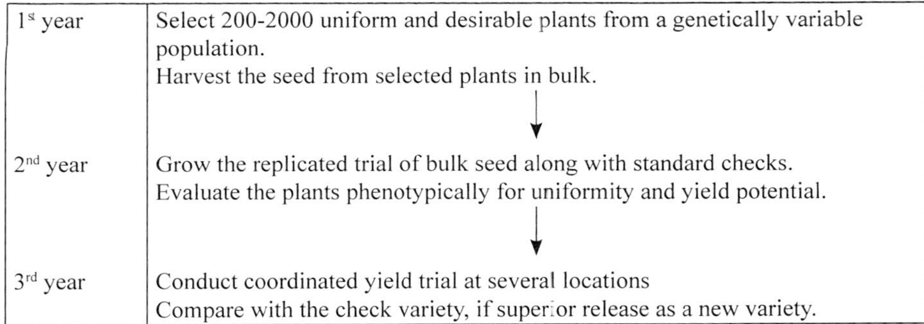

1st year	Select 200-2000 uniform and desirable plants from a genetically variable population. Harvest the seed from selected plants in bulk. ↓
2nd year	Grow the replicated trial of bulk seed along with standard checks. Evaluate the plants phenotypically for uniformity and yield potential. ↓
3rd year	Conduct coordinated yield trial at several locations Compare with the check variety, if superior release as a new variety.

Fig. 6.1: A generalized schematic diagram of mass selection method

Advantages and limitations

The major advantages and drawbacks of this method are as follows-

Advantages

1. Simple and quick method of crop improvement with single generation selection cycle.
2. As the end product of the selection is a mixture of several purelines it carries considerable genetic variation though they are phenotypically uniform.
3. Due to the presence of genetic variability the improved variety shows higher adaptability and buffering capacity in different environments.
4. Large populations can be handled.
5. It is an inexpensive method.

Limitations

1. As selection is based on phenotype, traits with low heritability cannot be improved. It is not effective for selection of the characters that are highly influenced by environmental factors.
2. Superiority of the selected plant is due to genetic cause or effect of environment cannot be confirmed.
3. It is not possible to differentiate between heterozygous and dominant homozygous individuals when dominance gene action is present. Selection of heterozygous results into segregation in subsequent generation which require another cycle of selection. So the additive gene action of target trait is desirable for mass selection.

4. It requires uniform environment for getting the best result of selection because the selection is based on phenotypic value.
5. Mass selected variety is not as uniform and potential as pureline variety.

Application of mass selection in spice breeding

Mass selection is a breeding method that has been in use from the very beginning of agriculture. Most of the present day's cultivated species are developed through the major contribution of mass selection. Though now a days the use of this method has been restricted in purification of varieties and production of nucleus seed till a number of seed spices variety have been developed through this method. In coriander, variety Guj.Cor.1, Co.1, Co.2, Co.3, Co(CR). 4, Guj. Cor.2, CS 287, Hisar Sugandh, HisarSurabhi have been selected from various local and indigenous germplasm. Similarly Sadhana has been developed from local Alur collection, Swathi from Nandyal germplasm, Sindhu mass selected from Warangal local, HisarAnand from Haryana collection, Azad Dhania-1 from Kalyanpur germplasm collection, Pant haritima from Pant dhania. Mass selected varieties have been developed from exotic germplasm also. Example- CIMPOS-33 from germplasm of Bulgaria and ACR-01-256 from Russia. Cumin varieties Mc. 43, Guj. Cumin 1, 5-404; fennel varieties Co.1, HisarSwarup, Azad Sanuf-1 are also developed through this selection method. Co.2, Lam sel.1, Pant Ragini, AM-01-35 are few example of mass selected fenugreek variety.

Pureline selection

Pureline is the progeny of a single, homozygous, self-fertilized plant. So they are group of individuals with identical genetic constitution and maintain their features in successive generations through selfing. In 1903 Johannsen developed the theory of pureline. Phenotypic expression of an individual is the result of its genotypic and environmental effect. As the pureline of a parent is genetically identical so the variation present within a pureline is not heritable but due to environment. Therefore, selection within pureline is not effective.

The pureline selection is the individual plant selection from a mixture of homozygous lines. Unlike mass selection, here the improved variety is not a population of mixture of several superior genotypes rather a single best genotype selected from the base population. The base population may be landrace, introductions, mass selected populations or segregating populations of a cross. The period of selection depends on the genetic constitution of the population. High level of homozygosity in landrace, mass selected populations leads to minimum number of selection cycle for the development of pureline. In case of segregating populations, more generations of selection is to be practiced for achieving desired level of homozygosity.

Pureline selection, like mass selection, exploit the existing variability of the target population, do not create new variation. This type of selection aims at selecting the best performing genotype from a population. So the maximum genetic improvement of a self-pollinated crop over base population without creating new variation in the base population can be achieved through pureline selection. But at the cost of maximum genetic improvement it narrows down the genetic base.

Steps in pureline selection

Year 1/Season 1: Space plant a variable population (landrace, segregating population of a cross) as base population. Select desirable plants for trait(s) of interest along with high yield. The number of plants to be selected depends on the genetic variability of the base population and the space, resource available for growing the individual progeny rows in the next generation. Selection of less number of individuals possess a risk of losing promising genotypes at very early stage which will not reappear again. Harvest the selected plants individually.

Year 2/Season 2: Grow single plant progeny rows of the selected individuals. Evaluate the progeny rows for desirable traits and residual heterosis within progenies. Discard the nonpromising and variable progenies. Harvest the selected progenies individually.

Year 3/Season 3: Evaluate the progenies in preliminary yield trial including the best variety of the crop as check. Select progenies on the basis of desirable characters, uniformity and disease reaction. Compare the yield potential of the selected individuals with check variety.

Year 4-6/Season 4-6: Conduct replicated yield trials at multi-locations (3-5 locations) for multiple years (2-3 years) to identify the best pureline in its area of adaptation. Observe the progenies for uniformity on the basis of flowering and maturity, disease resistance, yield, important quality parameters etc. best one or few lines are advanced to advanced yield trial.

Year 7/Season 7: Conduct advanced yield trial at multiple locations. The best available variety of the crop is used as check. Identify the highest yielding line and release it as a new pureline variety.

The procedure is given in Fig 6.2.

1st year — Obtain a variable population (mixture of purelines).
Space plant the seed.
Select 1000 phenotypically superior plants and harvest them individually.

↓

2nd year — Grow individual plant progeny row of selected plants.
Evaluate the plants within progeny for desirable traits. Discard the undesirable progenies.
Select 200 promising progenies and harvest them individually.

↓

3rd year — Grow the selected progeny in preliminary yield trial with suitable check.
Select 50 superior progeny with respect to desirable trait, uniformity and disease resistance.

↓

4th-6th year — Conduct replicated yield trial at several locations.
Select 10 progeny lines evaluating yield potential, maturity, quality traits and disease resistance.

↓

7th year — Conduct advanced yield trial with check variety.
Select the best performing line.
Release as a new cultivar after multiplication.

Fig. 6.2: A generalized schematic diagram of pureline selection method

Adantages

1. Identification of best pureline from a variable population confirms the maximum genetic advance over the base population.
2. Cultivar developed through pureline selection has greater acceptability by farmers and consumer due to its high uniform nature.
3. Traits of low heritability can also be improved as selection is made based on progeny performance.

Limitations

1. It requires more space and resources for progeny testing.
2. The improvement is limited to the isolation of the most desirable genotype from a mixed population.
3. The cultivar has a narrow genetic base due to genetic uniformity. This render them susceptible to disease pest and less adaptable to environmental fluctuations.

4. Discarding all the genetic variants except the best genotype leads to faster genetic erosion.

Application of pureline selection in spice breeding

This is the one of the most extensively used selection method in self-pollinated crop species. Many improved cultivars have been isolated from genetically variable local varieties. A large number of improved varieties of coriander, cumin, fennel, fenugreek have been developed through pureline selection.

Popularly cultivated coriander varieties RajendraSwathi selected from Muzaffarpur collection, DWA 3 selected from Karnataka collection; cumin variety Guj Cumin 2 selected from MC-43 through pureline selection. Guj Fennel 1 (selection from Vijaypur local), Pant Madhurika are some pureline fennel varieties. Some other example of varieties developed though this method are, fenugreek varieties Rajendrakanti (selection from Raghunathpur collection), RMt.1 (selection from Nagpur local), Hisar Sonali, Hisar Suvarna, Hisar Madhavi (selection from local germplasm), Hisar Muktha (selection from mutant line).

6.3.3 Pedigree method

At the beginning of crop improvement mass and pureline selection was the predominant methods for the development of new cultivar utilizing naturally occurring variation. As natural genetic variation got exhausted rapidly due to vigorous breeding, there was a call for creating new genetic variation for selection programme. Then hybridization was incorporated in the breeding scheme. Now the segregating population derived from the crossing between parents became the new base population for the crop improvement. Pedigree method is one of the most widely used method in self-pollinating crop which handle the segregating generation. This method aim to select a complete homozygous line from segregating population which will outperform both the parent used in hybridization.

The pedigree method was first described by H. H. Love in 1927. This method got its name from the word ‘pedigree’ means record of ancestry or lineage of an individual. Detail record of selected plants in each generation are documented by breeder which helps to trace back a plant’s ancestral relationship back from F2 generation. Pedigree breeding method consists of the following basic steps- crossing among selected parents, growing segregating generation, individual plant selection in subsequent generations. Selection is continued until acceptable level of homozygosity and homogeneousness is achieved.

Regarding the pedigree record keeping, record should be simple, informative, easy to understand and manage. The most common method is using numerical numbers serially to denote the number of cross, plants or family. For own convenience breeder can modify it accordingly and can put down some information regarding the selected individuals. It helps greatly in saving resources by avoiding the growing plants of same parentage repeatedly in advanced generations.

Steps in pedigree selection

The detailed procedure of pedigree breeding method is as follows-

Year 1/season 1: To initiate the breeding programme, selection of parents for hybridization is first step. Parents should be selected based on their average performance, adaptability and genetic divergence. Success of the breeding scheme to achieve a predetermined goal largely depends on the genetic distance between the parents. The chance of getting new gene combinations in F2 depends on the number of locus for which the parents differ from each other. So distantly related, homozygous parents are crossed to develop F1 seed.

Year 2 / Season 2: F1 seeds are space planted for allowing optimum growth of F1 plants to produce maximum F2 seeds. F1 plants should be closely observed for the traits for which the parents are differing . Plants resembling the female parent may be a product of selfing. Selfed plants should be rogued out. Genetic markers can assist a breeder to confirm the true hybridity.

Year 3/Season 3: individual plant selection starts from F2 generation. This is the generation with maximum genetic variation. Here each plant is different from others with respect to combinations of genes aroused due to meiotic recombination and random segregation. So it is desirable that F2 population should be as large as possible to accommodate all the segregants. 2000-5000 F2 plants are space planted to facilitate individual plant evaluation and selection. The best variety of the region is used as check and planted at regular interval throughout the F2 population for easy comparison. Individual plant selection is carried out with moderate selection intensity (about 10%). As there is considerable heterozygosity in F2 generation, a large number of F2 plants are selected and tested for their potentiality in subsequent generation. Genetic identity of each selected plant is documented. They are harvested individually.

Year 4/Season 4: The individual plant progeny rows of selected F2 plants are grown with check variety. Homozygosity and heterozygosity of the selected F2 plants can be confirmed seeing the segregation in the progeny row. The unpromising progenies compared to check variety are discarded. An outstanding plant in an otherwise rejected progeny row can also be advanced to next generation. Selected plants are harvested separately and their details are recorded.

Year 5/Season 5: Selected F3 plants are progeny rowed in F4 generation in the same way as in the previous generation. Now all the F4 progeny rows that are derived from the selected individuals of a same F3 progeny row constitute a family. As the generation advance variation within family decrease and between family increase. Some F4 families can be completely discarded based on their disease pest susceptibility or poor yield. Now selection is practiced among the progenies within a family. Superior plants from the progeny row are harvested. Selection of plants from more families is more rewarding than selecting more progenies from few families as the former represent more F2 plants in the next generation. Selection of many superior plants stemming from the same F2 plants is to be avoided using the pedigree record.

Year 6/Season 6: The selected plants are grown in plant-to-row fashion in a family with commercial seed rate. As variation within family is reduced to a significant extent, progeny row selection is practiced instead of single plant selection. Performance of families are evaluated with one or several check varieties that are locally adapted cultivars. The selected families are harvested in bulk to raise a sufficient amount of seeds for preliminary yield trial.

Year 7/Season 7: The lines selected in F5 generation are grown in a preliminary yield trial. These lines are replicated along with some check cultivars. A comprehensive evaluation of target traits to be improved, yield, disease resistance, quality parameters are performed. Top yielding families are retained for advance yield trial.

Year 8-10/Season 8-10: Superior experimental lines are again screened for two or more years in different locations under advance yield trial. The most outstanding experimental line is released as new cultivar.The procedure is given in Fig 6.3.

1st year	Select homozygous parents and cross them. Harvest F_1 seed. ↓
2nd year	Grow F_1 plants by space palnting. Harvest F_2 seed in bulk. ↓
3rd year	Grow 2000-5000 widely spaced F_2 plants. Select 200-500 promising plants based on phenotype. Harvest seed from each plant separately. Keep record of the selected plants. ↓
4th year	Grow plant to progeny rows. Select superior plants from superior progenies. Reject undesirable progenies. Harvest individual plants and keep record. ↓
5th year	Grow plant to progeny rows. Reject poor plants from progenies and reject poor families as a whole. Harvest selected plants individually and keep record. ↓
6th year	Grow plant to row in a family. Select superior progeny row as a whole. Discard poor families. Harvest the selected family in a bulk. ↓
7th year	Conduct preliminary yield trial. Evaluate the selected lines with check variety. Screening for yield, quality and disease resistance. Save the top yielding family. ↓
8th to 10th year	Conduct advanced yield trial at several locations. Release the most promising line as new variety.

Fig. 6.3: A generalized schematic diagram of pedigree method of selection.

Advantages and limitations

Advantages

1. Pedigree record keep the complete information about the line of descent and useful key features of the selected individuals.

2. Pedigree records helps breeders to advance only those genotypes which have the desirable gene combinations.
3. Elimination of inferior genotypes well before replicated yield trial saves resources.
4. Genotypic superiority of the selects are confirmed by practicing progeny row selection.
5. It provides the maximum opportunity to a breeder to show his skill in selecting superior plants.

Limitations

1. Record keeping is a laborious and tedious job.
2. Space planting for individual plant selection and growing more progeny rows demands for more resources (space and labour).
3. It is not suitable for the plant species where characterization of individual plants are possible.
4. It requires 10-12 years for developing a cultivar.
5. Selection for quantitatively inherited traits (yield) in early generations (F2 or F3) is ineffective in most of the cases.

6.3.4 Bulk population breeding

In pedigree method, we have already seen that plants are artificially selected from very first segregating generation (F_2) and pedigree records of the selects are maintained to track the line of ancestry. But this is practically not feasible when a large number of crosses are to be handled simultaneously. So the perpetuation of segregating generations as a bulked pool of genotypes of without artificial selection especially in the first few generations is an alternative to the pedigree method. H. Nilsson-Ehle first used bulk method for improvement of winter hardiness in wheat.

The salient features of bulk population method are:

1. Segregating generations are advanced by bulking the harvest. Individual plant selection starts in later generation when plants become considerable homozygous. Generally single plant selection starts at F_5 or F_6. In F_6 generation most of the plants of self-pollinated crops attain nearly 98% homozygosity. So selection and evaluation among these true breeding types are more effective.

2. Natural selection is the only selection force acts on the segregating generations until favourable situations prevail for starting artificial selection. Bulk breeding method is also termed as evolutionary breeding method as natural selection plays a pivotal role in changing the genotypic composition of the population.
3. Natural selection improves the population with respect to adaptability. If a population is subjected to biotic or abiotic stresses like disease-pest infestation, salinity, drought, high or low temperature, water logging, change in photoperiod etc., the more competitive and better adapted genotypes will survive and successfully reproduce to advance the generations.
4. Artificial selection may be practiced along with natural selection to accelerate the breeding programme. The direction of artificial selection must be the same that of natural selection only to assist the process.
5. This method is suitable for close spacing crop species (mostly cereals and pulses) where observing individual plant is not possible and interplant competition is present in thick stand. In most of the horticultural crops, where much longer spacing is maintained and interplant competition is not desirable, bulk breeding is not advantageous.

Procedure

The basic procedure of bulk method to isolate purelines from segregating populations are described below. Though the duration of bulking depends on the objective of breeding and environmental effects.

1st year (Hybridization): desirable plants based on the breeding objective are selected and sufficient crosses among them are made.

2nd year (F1 generation): F1 plants are space planted for obtaining maximum F2 seeds. True hybridity of the F1 plants are checked. Selfed plants are rogued out. F1s are harvested as bulk.

3rd year (F2 generation): F2 seeds are planted in a bulk plot at commercial seed rate or even in thicker stand. All F2 plants are bulk harvested.

4th to 7th year (F3 to F6 generation): A random sample of composite F2 seeds are sown in bulk plot. Close spacing is maintained to create competitive environment and to accommodate maximum plants in a specified area. Harvests are composited to grow the next season. The same procedure is followed upto F5 or F6 generations till considerable level of homozygosity is achieved. During this period generally artificial selection is not practiced. Only natural selection acts on the population to eliminate the ill-adapted plants and increase the frequency of desirable genotypes.

8th year (F7 generation): as sufficient level of heterozygosity is achieved at this stage, single plant selection may be initiated. Plants are grown in larger spacing in this generation for better expression of individual plants. Plants with superior phenotype are selected and harvested individually.

9th year (F8 generation): progeny rows of each selected individuals are grown. At this stage most of the progenies are homozygous. So promising progenies based on visual evaluation are harvested separately. Progenies with obvious defects and segregating progenies are discarded. But segregating progenies with potentially superior individuals can be advanced further.

10th year (F9 generation): selects of the previous generation are grown in multi row plot preliminary yield trial with best commercial check variety. Yield and other quality parameters are compared with the check. Lines are also evaluated for major disease and pest resistance.

11th to 13th year (F10 to F12 generation): advanced yield trials are conducted at multiple locations over the years. Diligent testing of the genotypes for yield, quality and resistance are done. Best performing line compared to standard check is multiplied and released as new cultivar.

Modification of bulk breeding

In accordance with the need of breeding programme several modifications have been done in different steps.

- Artificial selections may be started in F_3 or F_4 generation to fasten the process of increasing the frequency of desirable genotypes in the population.
- To avoid the effect of random sampling in the early segregating generations single seed from each individuals can be used to advance the next generation without performing any selection. This modification reduces the risk of random drift. This process is most popularly called single seed descent method of breeding.The procedure is given in Fig 6.4.

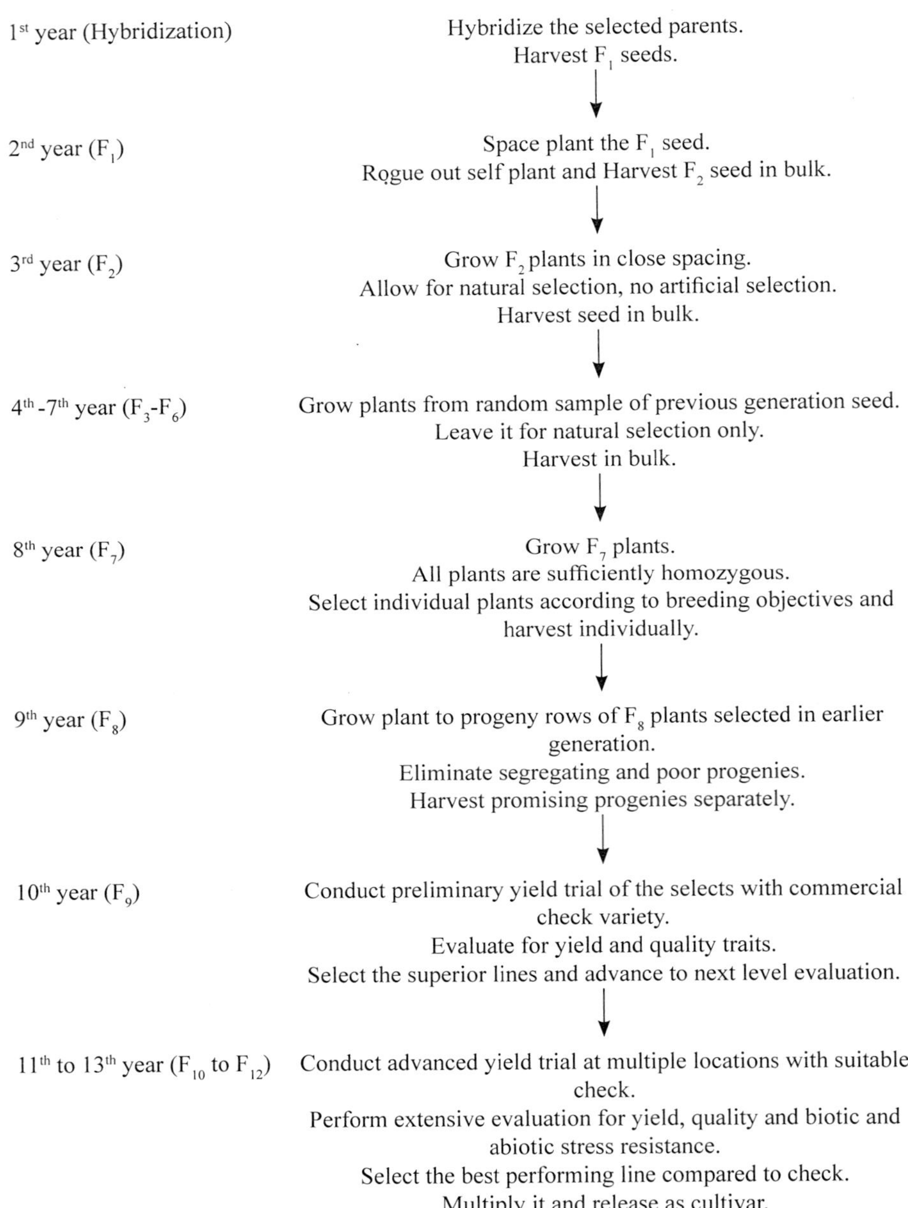

Fig. 6.4: A generalized schematic diagram of bulk population method.

Advantages and limitations

Advantages

1. This is a comparatively easy and inexpensive method.
2. Large number of segregating populations can be handled with minimum resources.
3. Individual plant selection at later generation when plants attain considerable homozygosity eliminate the risk of deceptive nature of heterozygous plants.
4. Utilize the advantage of natural selection to eliminate the undesirable less adapted genotypes.

Limitations

1. It demands considerably much longer time.
2. Natural selection may eliminate some agronomically superior segregants when there is a negative association between agronomically desirable trait and fitness under natural selection.
3. Sampling during growing large segregating population, sample size or sampling procedure may not form a sample which truly represent the whole population. This may cause loss of genetic variation and it results an undesirable change in the frequency of genotype. This is random drift.
4. For advancing generations, off season nursery or green house cannot be used as environment of the off season may be different from that of main season.

6.3.5 Backcross breeding

Backcross is the crossing of F_1 to either of its parents. It is a special breeding method used to improve one or two specific trait in an otherwise well-established high yielding cultivar. Unlike other breeding methods of self and cross-pollinated crops, where identification and isolation of a breeding line with entirely new genetic makeup from the base population is expected, here an existing cultivar is improved by adding few genes without disturbing the original genetic background. Backcross breeding is quite popular in animal breeding but use of it in plant improvement was first suggested by Harlan and Pope (1922). The basic procedure entails crossing a popular adapted variety deficient for particular trait with a donor parent having specific trait to be transferred and repeated crossing of F_1 with the popular cultivar to retrieve plants similar to the recurrent parent with the addition of gene under transfer.

Pre-requisite for backcross breeding

1. A suitable recurrent parent variety which requires improvement in one or two qualitatively inherited character or quantitative character with high heritability.
2. A donor parent which is the source of genes missing in recurrent parent.
3. The character to be transferred must have high heritability and must have high expressivity in the genetic background of the recurrent parent.
4. A sufficient number of back crosses should be made so that the genotype of recurrent parent is recovered in full.

Procedure

The general procedure of backcross method involves identification of recurrent and donor parents, F1 of them is crossed to recurrent parent. In each backcross generation plants with desirable characters are selected and crossed to recurrent parent. Repetition of this process is followed for four to six generations. The duration of backcross breeding method depends on various factors like number of genes controlling the character under transfer and the nature of gene action (dominance or epistatic), effect of environment on expressing the character etc.

Transfer of a single dominant gene

Let us consider, a popular cultivar is susceptible for a disease. Susceptibility for the disease is governed by a single recessive gene (r). Now a breeder is intended to improve this cultivar by transferring resistant dominant gene (R) from another source (donor parent), the steps are to be followed:

Season 1: Recurrent parent (rr) is crossed with donor one (RR) to produce F1 (Rr), where recurrent parent is usually used as female and donor parent as male.

Season 2: F1 plants are grown and backcrossed with recurrent parent to obtain first backcross (BC1). Enough backcross should be made to produce sufficient BC1 seed as only 50% of the total BC1 plants will be resistant and to be used for further crossing.

Season 3: BC1 plants with genetic constitution Rr and rr are grown. Plants are evaluated for disease reactions by creating epiphytotic condition. Resistant plants (Rr) are selected and crossed with recurrent parent (rr) to produce BC2.

Season 4: BC2 plants are grown and again they would segregate as 50% resistant (Rr) and 50% susceptible (rr). Resistant plants are crossed with recurrent parent and produce BC3 seed.

Season 5: BC3 seeds are sown and BC4 seeds are produced the same way as BC3. In this generation beside selection for resistance a sight is to be kept on the recurrence of the genome of the recurrent parent. Most of the plant will resemble the recurrent parent for most of the trait.

Season 6: As previous BC4 plants are screened for disease resistance. The resistant plants (Rr) are self-pollinated to produce BC4F2 seeds.

Season 7: the genetic constitution of BC4 plants that are selected is Rr. So upon selfing they will produce three types of genotypes: RR, Rr and rr. The BC4F2 generation will consists of segregants of resistant (RR, Rr) and susceptible (rr) plants. The susceptible plants are discarded and remaining resistant plants are selfed. Selfed seeds (BC4F3) from the resistant plants are harvested individually.

Season 8: Individual plant progeny rows (BC4F3) from self-pollinated plants are grown separately to differentiate between RR and Rr. Progenies of RR plant will be uniform for resistance and that of Rr plant will be segregating for resistance. Segregating progenies are rejected. Uniform and resistant progenies are harversted and bulked together which will represent the improved version of recurrent parent.

Transfer of a single recessive gene

If the desired trait under transfer is controlled by recessive gene then an additional step of selfing is to be carried out for identifying the homozygous recessive genotype that is to be backcrossed with recurrent parent. During recessive gene transfer, in BC1 generation which genotype among the two (RR and Rr) carries the desired recessive allele (r) cannot be determined phenotypically. So it is necessary to conduct a progeny test by growing the selfed seeds of the two genotypes (RR and Rr). From the selfed generation homozygous resistant plants (rr) are identified and crossed with recurrent parent for next backcross generation.The procedure is given in Fig 6.5.

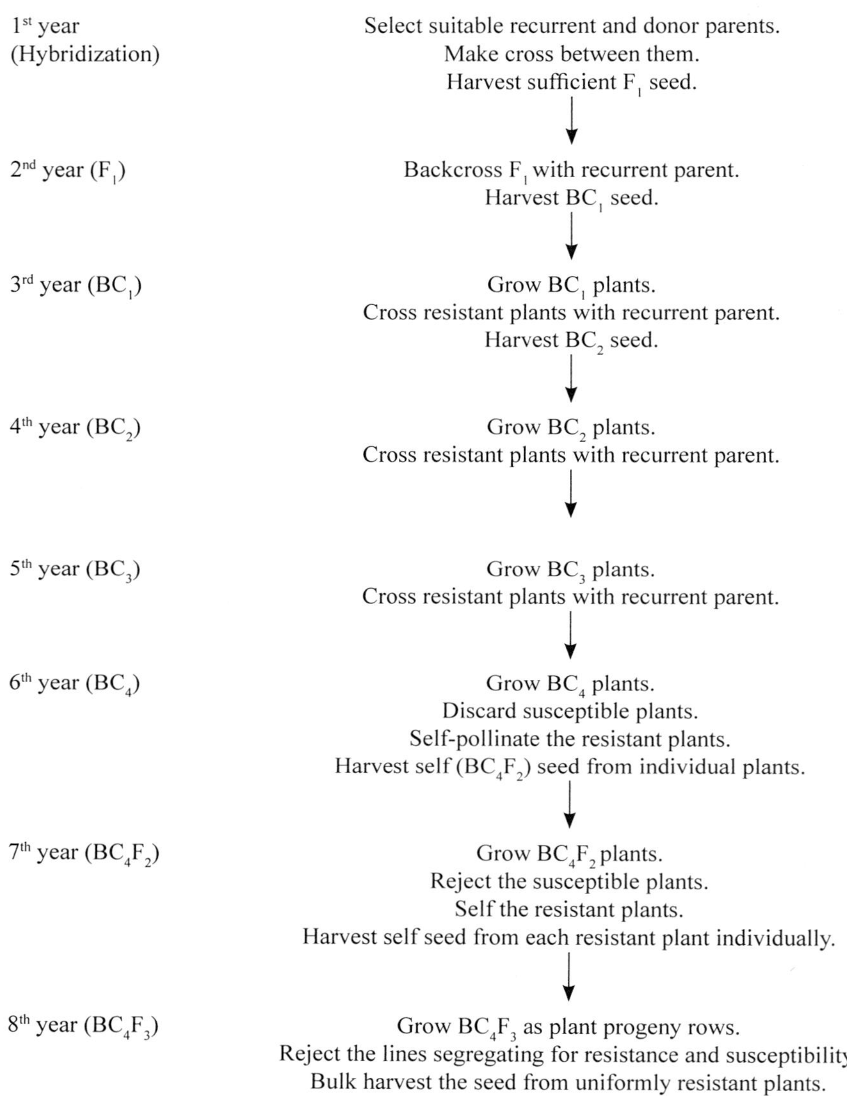

Fig. 6.5: A generalized schematic diagram of backcross breeding method.

Advantages and limitations

Advantages

1. This method is repeatable. Breeder can repeat the process if same backcrossed cultivar is needed.
2. Release of new cultivar do not require extensive field trial as adaptability of the cultivar in that particular area is already known.

3. Off-season nursery can be used to advance the generation.
4. Genes can be introgressed from unadapted germplasm also. It allows interspecific gene transfer.

Limitations

1. The presence of undesirable linkage may hinder the process of improvement.
2. Backcross breeding is not effective for transferring quantitative traits and traits with low heritability.
3. This method does not lead to an improved cultivar superior to the recurrent parent, except for the character that is transferred.

6.4 Breeding Methods for Cross Pollinated Crops

The difference in the selection methods between self and cross fertilized species is largely due to the difference in their genetic constituents. As cross pollinated crops are highly heterozygous, breeding cross pollinated species aims at improving population rather than the improvement of individual plants, as practiced in self-pollinated species.

The two major breeding methods of cross-pollinated species are- i) Heterosis breeding/ hybrid breeding and ii) Population improvement. In both the cases heterozygote advantage is utilized either in the form of identifying the most productive heterozygote from a population (heterosis breeding) or by accumulating favourable alleles in the population through repeated cycles of mating and selection (population improvement). But they have difference in the gene effects which they exploit in the process of improvement. Hybrid breeding utilizes dominance gene effects whereas population improvement relies largely on additive gene effects.

6.4.1 Heterosis breeding

A hybrid cultivar is the first generation offspring (F_1) of planned crosses between genetically unrelated parents. These parents may be pureline, inbred, varieties or populations. Luxurious growth and yield of a hybrid cultivar is due to the phenomenon heterosis. The genetic basis of heterosis reveals that the non-additive gene action of the trait(s) of interest contribute to the superiority of F_1 progeny over its parents. Hybrid cultivar is more popular in cross-pollinated crops as there is a preponderance of non-additive gene action for most of the economically important traits in allogamous crops. But hybrid cultivar can be developed in autogamous crops also where heterosis is commercially exploitable.

Hybrid breeding involves three basic steps:

A) Development of inbred lines

B) Evaluation of inbred lines

C) Production of hybrid by combining inbred lines

Development of inbred lines

Inbred are near homozygous genotype of an open pollinated variety derived from heterozygous population through continuous forced selfing (inbreeding). Generally inbred lines are developed from open pollinated population, improved population of cross-pollinated species, F_2 progeny derived from single, three way cross or any other heterozygous population. Superiority of F_1 depends upon the performance of inbred lines and extraction of elite inbreds is dependent on the source population. So selection of source population with broad genetic base is crucial.

Standard method of inbred development: A number of desirable plants are visually selected from a source population and selfing is carried out individually. In the following season, the seeds of the selected plants are sown in ear-to-row and selection is practiced within and among the progenies along with inbreeding the selects. The vigorous plants are advanced to next inbreeding cycle. Continuous selfing of a heterozygous open pollinated population splits into an array of homozygous lines. With successive generations of inbreeding, the variance between lines increases and within line decreases. Four to five generations of selfing will results in satisfactory level of homozygosity to be used as an inbred.

Early testing: It has been seen that, among several hundreds of inbreds only a few combine well with others to produce promising hybrids. Therefore evaluation of lines after development of inbreds demands a considerable amount of resources and time. So a process has been devised where general combining ability of inbreds is tested at an early stage of inbreeding. In this method plants from early generations of selfing (S_0, S_1 or S_2) are selected based on their phenotype and they are selfed. Simultaneously these plants are test crossed with a broad genetic base tester. On the basis of replicated performance of test cross progenies, selfed lines are selected and retained for further inbreeding. This is repeated in every generation of inbreeding and finally a manageable number of good combiner inbreds are developed for further evaluation. The only drawback of this method is that some potentially good combiners with agronomically undesirable characters are eliminated during inbreeding.

Evaluation of inbred lines

The worth of an inbred line is determined by its ability to combine well with other inbreds and its genetic contribution in hybrid combinations, not in its production potential. If there are n inbreds, n(n-1)/2 numbers of crosses are to be made for testing the performance of inbreds in all possible combinations. Testing of inbreds are done in a step-wise manner starting from phenotypic evaluation to general combining ability test through top cross and specific combining ability test by single cross evaluation.

Phenotypic evaluation: Inbreds are screened preliminarily based on their performance in replicated trial. Though it is not compulsory an inbred good combiner must be good yielder also. Still to maintain and multiply the inbred as a parent of an elite cross, it must be vigorous and free from diseases.

Evaluation for General Combining Ability (Topcross evaluation): General Combining Ability is the average performance of a genotype in a series of cross combinations. GCA is a function of additive portion of the genetic variation. Evaluation of GCA of an inbred is done by crossing it with a series of other inbred lines and comparing the average performance of the single cross progenies. Top cross provides an easiest mean of isolating good general combiners. Here inbred lines are crossed with a broad genetic base tester (open pollinated cultivar, double cross population or synthetic). The test cross progenies are compared in performance trial and based on this about 50% inferior combiners are eliminated.

Evaluation for Specific Combining Ability (Single cross evaluation): In the final step of the inbred evaluation, the lines selected as good general combiners are crossed in all possible single crosses. Single cross provides a reliable estimate of specific combining ability (SCA). From SCA analysis, the relative abundance of non-additive genetic variance (dominance and epistatic) can be predicted. Non-additive gene action, specially additive action, is largely responsible for the excellence in yield performance of hybrid. Usually, a diallel system of mating is adopted to cross each inbred with every other one. If there are n inbreds, n(n-1)/2 single crosses (excluding reciprocals) are evaluated in replicated trials. The best cross selected from all possible single crosses is the most promising single cross hybrid expected to utilize maximum amount of non-additive variance.

Thus evaluation of inbreds for GCA followed by SCA aims at utilizing both additive and non-additive variance in hybrid progeny.

6.4.2 Production of hybrid by combining inbred lines

Hybrid is produced by crossing two genetically unrelated inbreds where one is used as male and other as female. But bisexual nature of flower and monoecious plants in most of the crops call for techniques to adopt for successful crossing. All these are not equally applicable, depends on crop species, economic feasibility etc.

Hand emasculation and pollination: this is the process of eliminating the male reproductive organ or male inflorescence manually. This is laborious and expensive which makes the cost of hybrid seed higher. It is feasible where large number of seeds are obtained from emasculating and pollinating a single flower.

Male sterility: This is a potential aid to commercial hybrid seed production which reduces the cost of hybrid seed by reducing the labour engaged in manual emasculation. Male sterility is a genetic emasculation technique where presence of male sterility gene in cytoplasm or nucleus results in the production of non-functional pollen. This is generally controlled by single recessive gene (*ms*). In genetic male sterility (GMS) system, *ms* gene is present in nucleus. Male sterile plant (*msms*) is crossed with male fertile line (*MsMs*) to produce male fertile F1 (*Msms*).

3 line hybrid system: cytoplasmic genetic male sterility (CGMS) is the most commonly used sterility system used in commercial hybrid production of the most of the crop. The expression of fertility/sterility is governed by the interaction of male sterile (*ms*) gene present in cytoplasm and restorer gene (*R*) in nucleus. Restorer gene (*R*), in its dominant form, has the ability to restore the male fertility in male sterile lines. In this system, 3 types of lines- A line (*msms/rr*), B line (*MsMs/rr*) and R line (*MsMs/RR*). A line is the male sterile line, used as female parent in hybridization. B line is isogenic to A line, with normal instead of sterile cytoplasm. A line is maintained by pollination from B line in isolation, that's why B line is also called as maintainer line. R line is the pollen parent of the hybrid used to pollinate the A line and restore fertility of hybrid seed.

The success of hybrid production using CGMS depends on the stable expression of male sterility and availability of restorer genes.

Self-incompatibility: Another tool of hybrid breeding where functional normal pollen cannot participate in self-pollination due to some physiological causes. Cross compatible best combiners can be crossed to obtain F_1 seed in both parents. Though this method is not common in use.The procedure is given in Fig 6.6.

Production of Inbred lines	Select genetically diverse open-pollinated source population. Self or close inbred selected superior plants for 6-7 generations. Isolate nearly homozygous inbred lines.
	↓
Evaluation of Inbred lines	Phenotypic evaluation: select better performing inbreds from replicated trial. Evaluate phenotypically for agronomic superiority.
	↓
	Top cross test: test the phenotypically superior inbreds for general combining ability (GCA). Cross the selected inbreds with a broad genetic base tester. Compare the top cross progeny performance and select inbreds based on the performance.
	↓
	Single cross test: cross the inbreds with good GCA in all possible combinations. Test the inbreds in half diallel mating design. Evaluate the single cross progeny performance and select best cross combination based on specific combining ability (SCA) value.
	↓
Hybrid production	Cross two or more inbreds with wide genetic diversity in specific manner to produce single cross, double cross or polycross.

Fig. 6.6: A generalized schematic diagram of heterosis breeding.

Application of heterosis breeding in spice improvement

Hybridization is one of the most frequently used plant breeding method. It has been extensively used in improving yield, quality and biotic and abiotic stress resistance in many spice crops. Intergeneric and inter-specific crosses have also shown tremendous potential in improvement of specially disease resistance and introgression of quality traits. Black pepper variety Panniyur 1 and Panniyur 3 are selection of a cross between Uthirankotta and Cheriyakaniyakadan. Several inter-varietal hybrid of pepper are there with high yield and *Phytophthora* resistance. Interspecific hybrid between *P. nigrum* X *P. barberi* and intergeneric crosses with *Ammomum subalutum, Hedychium coronarium*have been attempted for introducing katte resistance.

6.4.2 Population improvement programmes

A population is defined as a large group of individuals of same species capable of intermating among themselves. A population of open pollinated variety is characterized by the presence of heterozygous and heterogeneous group of individuals where each individual is different from others with respect to different gene combinations. Sprague (1966) described population improvement as systematic technique to improve the mean performance of a random mating population by eliminating less productive allele through repeated cycles of selection. Open pollination leads to appearance of new genotypes in the selection cycle which were not present in the base population. Thus it maintain a genetic variation and provide an opportunity to breeder for further improvement in future.

A random mating population maintain its gene and genotype frequency unchanged generation after generation provided there is no selection. This is called Hardy-Weinberg equilibrium. So selection for desirable plants in a random mating population dislodge its position from the equilibrium and random mating among the individuals shift the population to a new equilibrium position with improved mean performance. Continuous accumulation of additive gene effects and production of new gene combinations after each random mating never result this improvement to reach a dead end.

Concept of recurrent selection

Recurrent selection is reselection generation after generation with interbreeding of selects to provide for genetic recombination.

A recurrent selection is consists of three main steps:

1. Phenotypically superior individuals are selected from population.
2. Selfed progeny of the selected individuals are tested for yield and other characters.
3. Interbreeding among the individuals whose selfed progeny showed better performance.

The population derived from the intercrossing of superior individuals is the improved population and it serves as the base population for the next selection cycle. This cycle of selection is repeated several times. The improvement of the population through recurrent selection depends on the genetic variation present in the base population.

The advantage of the recurrent selection is that the ceiling of improvement is not set by a single genotype, but by the most favourable gene combinations present in a group of foundation plants. So the chance of evolving superior recombinants through recombination provide for effective selection for longer period.

The population improvement methods are generally classified into two categories based on the types of population improved:

i) Intra-population improvement methods

ii) Inter-population improvement methods

Intra-population improvement methods

The selection is practiced within a population. The end product is elite inbred lines or population. Different types of intra-population improvement methods are-

a. Mass selection

b. Half-sib family selection

c. Full-sib family selection

d. Selfed family selection

Inter-population improvement methods

Here simultaneous improvement of two complementary populations occurs. The selection is based on the cross performance between two populations. As a result it exploits interpopulationheterosis. The most commonly used method of inter-population improvement is reciprocal recurrent selection.

Mass selection

Mass selection is the oldest and simplest method of crop improvement for cross pollinated species. As the selection is solely on phenotypic performance, it is also called phenotypic recurrent selection. The unit of selection is individual plant.

Procedure

Year 1/Season 1: Desirable plants are selected from source population based on their phenotypic superiority. The source population which is to be improved may be local variety or bulk population. Undesirable plants are rogued out before flowering to eliminate their contribution in seed development of selected individuals. Open pollinated seeds from the selected plants are harvested. Equal amounts of seeds from each selected individuals are bulked. This is the completion of a selection cycle. This cycle may be repeated till target level of genetic gain and homogeneity is achieved. After attaining homogeneity preliminary and advanced yield trials are conducted using the original base population as check to gauge the improvement.

Advantages

1. Simple method with shortest selection cycle (one year).
2. It is an efficient method for improving trait with high heritability.

Limitations

1. No genetic basis of selection as progeny testing is not conducted.
2. No control over pollination. Pollen from undesirable plants also take part in the pollination of selected plants.
3. No estimation of G X E interaction.

Application of mass selection in spice breeding

Mass selection has contributed a significant number of varieties in spice crops. Widely cultivated black pepper varieties Panniyur 2, Panniyur 5, Panniyur 7, IISR Sakthi,; cardamom varieties PV 1, PV 2, ICRI 1, ICRI 3, IISR Suvasini, IISR Avinash; ginger varieties IISR Varada, IISR Mahima, IISR Rejatha; Turmeric varieties Rajendra Sonia, Megha turmeric 1, Suvarna, Suguna, Sudarsana, IISR Prabha, IISR Pratibha, IISR Alleppey Supreme; Cinnamon varieties RRL(B) C-6, IISR Navashree are some example.

Half-sib family selection

The unit of selection here is a family rather than individual plant like mass selection. The basic steps of family selection are-

1. Creation of families
2. Evaluation and selection of families
3. Recombination of selected families or plants to develop a new base population for next selection cycle.

Half sibs are group of individuals having one parent common. In plants, half-sib families are produced by growing progenies from selected female plants where the source of pollen pollinating the selected female plants are unknown. The source of pollen may be the population itself from where the females were selected, inbred line or synthetic variety.

Procedure

Year 1/Season 1: Heterozygous source population is grown. The desirable plants (S_0) are selected based on their phenotypes. Open pollination is allowed in the population and seeds from the selected plants are harvested individually. Remnant seeds of selected plants are kept for growing in future.

Year 2/Season 2: Half-sib progenies from the selected plants are grown for evaluating yielding ability and breeding value of the traits under improvement. The most productive progenies are selected and their seeds are composited for growing progenies in the next cycle.

Year 3/Season 3: The seed of selected superior progenies in second season are bulked or the remnant seed of S_0 plants whose progenies have been selected in second season are bulked. Bulked seed are grown in isolation (crossing block) for random mating. The composited open pollinated seeds harvested from the crossing block is used as the base population for the next selection cycle. This selection cycle of three years may be repeated as per the breeding objective. The procedure is given in Fig 6.7.

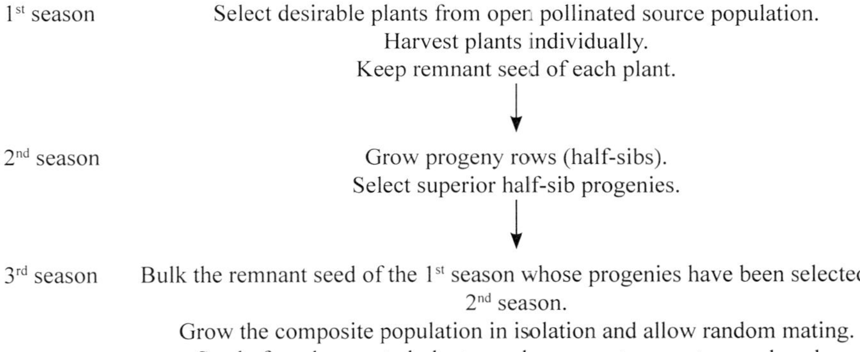

Fig 6.7: A generalized schematic diagram of half-sib (HS) family selection.

Full-sib family selection

Full-sib is defined as the family of individuals having both parents common. Full-sibs are generated by making crosses (sibbing) between phenotypically desirable parents from the base population (biparental mating).

Procedure

Year 1/Season 1: Random pairs of plants are selected based on phenotype from the base population and they are crossed to make full-sib families. Pair plants are individually harvested. Remnant seed of each full-sib cross is saved.

Year 2/Season 2: Full-sib progenies of selected pairs are evaluated in replicated yield trial and disease-pest nursery. Best families are selected on the basis of yield and disease-pest resistance.

Year 3/Season 3: The remnant seed of the full-sib crosses selected based on their progeny performances are mixed and grown in isolation. They are random mated for recombination in all possible combinations. The open pollinated seeds from the recombination block form a new set of source population for selection. The same cycle is repeated.The procedure is given in Fig 6.8.

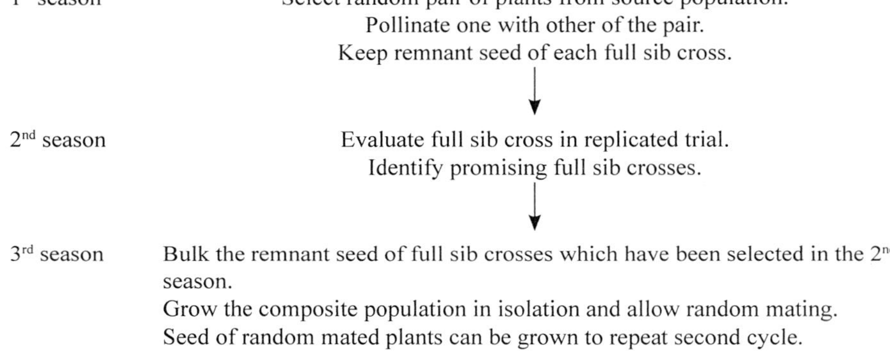

Fig 6.8: A generalized schematic diagram of full-sib (FS) family selection

Selfed family selection

Unlike other family selection methods, in this method promising plants of the source population are selfed. Forced selfing in all the cross pollinated crops are not amenable due to floral size, structure, self-incompatibility reaction etc. so it is applicable for those species only where selfing is practically feasible. The progeny derived from one generation of selfing is denoted as S_1 and S_2 for two generations of selfing.

The basic concept behind growing selfed progeny is that many deleterious recessive alleles are masked under heterozygous condition which are exposed and can easily be discarded in selfed generations. Accumulation of increased magnitude of additive genetic variation in the selfed progenies is another reason. Thus this method is very effective for increasing the frequency of desirable alleles with additive effects in the population.

Procedure

Year 1/Season 1: Phenotypically desirable S_0 plants are selected from the source population. Selected plants are self-pollinated. Selfed seeds (S_1) are harvested. A part of the selfed seeds of each selected individual are kept for evaluating in the next season and rest are saved as remnant.

Year 2/Season 2: Selfed progenies (S_1) are evaluated in replicated trial. Superior S_1 progenies are identified.

Year 3/Season 3: Remnant seeds of the best parents (decided on the basis of S_1 progeny performance) are composited and allowed for recombination in isolation. Thus the initial selection cycle is completed. The seeds derived from random mating initiate a new selection cycle.The procedure is given in Fig 6.9.

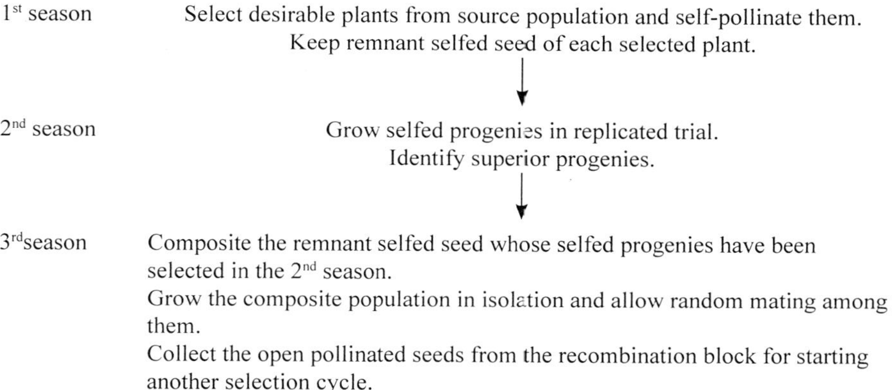

Fig. 6.9: A generalized schematic diagram of selfed family selection

Reciprocal recurrent selection

Reciprocal recurrent selection is an inter-population improvement approach where two heterogeneous populations are improved simultaneously with a purpose of improving the cross performance between two populations through exploitation of inter-population heterosis. Two populations are complementary, each one is used as tester for the other. This method is effective for improvement of general as well as specific combining ability of the population.

Procedure

Let us consider two populations namely A_0 and B_0 are used as complementary populations for improvement through reciprocal recurrent selection.

Year 1/Season 1: Desirable plants from population A_0 are selected and selfed. Simultaneously these plants (used as males) are crossed with a number of plants from population B_0 (used as females) to produce half-sib families of A_0 plants. Similarly selected plants of B_0 population are selfed and they are crossed as testers with a number of plants from A_0 as well to produce half-sib families of B_0 plants.

Year 2/Season 2: Half-sib families of selected A_0 and B_0 plants are evaluated in replicated yield trials separately. Superior progenies are identified based on their mean performance.

Year 3/Season 3: The selfed seed (in season 1) of the plants from populations A_0 and B_0 corresponding to the selected half-sib families (in season 2) are planted in separate crossing blocks. All possible intercrosses are made within each block. At harvest, two seed bulks, one from intercrossing of A_0 block and another from B_0 crossing block, are harvested separately. Now the new bulks are designated as A_1 and B_1 which are improved form of A_0 and B_0 respectively. This completes the first cycle of recurrent selection. According to the requirement, this cycle can be repeated using A_1 and B_1 as new source population. The procedure is given in Fig 6.10.

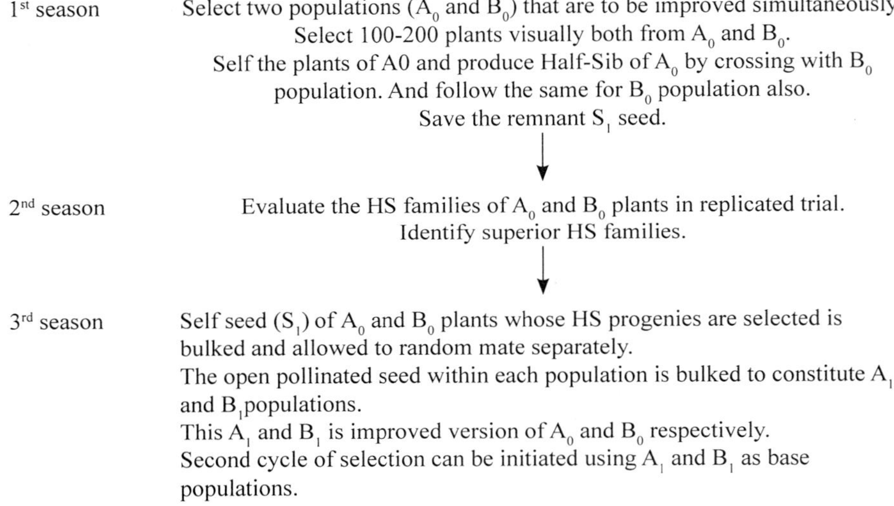

Fig. 6.10: A generalized schematic diagram of reciprocal recurrent selection

Application of recurrent selection in spice breeding

A few achievements in developing cultivar through recurrent selections are coriander varieties RCr 20, RCr 41, RCr 436, RCr 446 developed by recurrent half sib selection, RCr 435 developed by recurrent selection from local germplasm Jalore. Cumin varieties RZ-19, Guj. Cumin 3, RZ 19, RZ 209; fennel varieties RF 101, RF 125, RF 143, AF-01-119, fenugreek variety GujMethi 1 are some other example of varieties developed through recurrent selection.

6.5 Breeding of Asexually Propagated Crops

Crop plants, based on their mode of reproduction, are broadly categorized into two groups: sexually reproduced and asexually reproduced plants. Asexual reproduction is the mode of perpetuation of individuals without involving fusion of sexual gametes. This takes place either through specialized somatic tissue or by seeds developed without fertilization. A large number of horticultural crops are propagated through vegetative parts like- rhizomes (ginger), corms (gladiolus), tubers (potato), bulbs (onion), runners (strawberry), roots (poplar), stem cuttings (black pepper), leaves (Bryophyllum), suckers (banana), bulbils (lili) etc. These are also referred as clonal crops and group of individuals produced from a single plant through asexual reproduction are called clone.

Characteristics of asexually propagated crops

1. Asexual reproduction takes place by vegetative means through mitotic cell division. So genetic recombination and segregation do not occur in clones. Clones are genetically identical to their mother plants. Their genetic integrity can be maintained indefinitely over generations i.e., they are immortal.
2. They are mainly cross-pollinated crops.
3. They are highly heterozygous in nature due to gradual accumulation of natural mutations. High degree of inbreeding depression results on selfing.
4. Bypassing sexual reproduction leads to reduced flowering and seed setting, increased sterility.
5. Majority of these are perennials.

Breeding approaches

Absence of sexual recombination in vegetatively propagated crops provides some unique opportunities in the process of crop improvement programme. Any promising plant can be multiplied without segregation in successive generations irrespective of its genetic make-up (homozygous or heterozygous). But accumulation of systemic diseases (virus, mycoplasma, bacteria) and deleterious alleles in the absence of meiosis, various degrees of sterility are major disadvantages. Practice of selection within clone is ineffective as they have no genetic variation. With time, a clonal variety also shows genetic variation which is due to spontaneous mutation, accidental mechanical mixture or occasional sexual reproduction.

The basic procedures applied in the breeding of vegetatively propagated crops are generally same as that of sexually reproduced crops. The only difference is handling the segregating generations with minor modifications varies with the end use of the crop, extent of sterility and ploidy level of the species.

Asexually propagated species based on their dependence on asexual reproduction for maintenance are grouped into obligate and facultative asexually propagated species. Obligate asexually propagated species either produce sterile flower or do not flower at all. So vegetative propagation is the only choice. Their improvement is completely based on the exploitation of variation already present in the natural population or that created by spontaneous or induced mutations. On the other hand, facultatives have the capability to reproduce sexually beside asexual mode. They are in an advantageous position to utilize both sexual and asexual mode of reproduction. Here the breeding procedure aims at creating variations through meiotic recombinations by hybridization and maintaining the heterotic heterozygotes indefinitely by vegetative propagation. So the breeding procedure for clonally propagated species are clonal selection of natural variable population and hybridization followed by selection.

6.5.1 Clonal Selection

The selection procedure used for clonal crops are broadly termed as clonal selection. As all the individuals within a clone are genetically identical so selection cannot be practiced within a clone. Phenotypic variation within a clone is only due to environment, has no genetic basis. Therefore, clonal selection can be practiced in a genetically variable population where mixture of several clones are there. The improvement is limited to the isolation of the best genotype as in case of pureline selection in self-pollinated crops.

Selection procedure

1st year: Genetically variable clonal population is assembled. The origin of variation may be in the form of mechanical mixture of clones, bud mutations or occasional sexual reproduction within a clone. Superior clones with desirable trait(s), disease resistance and better adaptability in case of introduction to a new area are selected. Low selection pressure is maintained for the traits with low heritability. Propagule of the selected plants are harvested individually.

2nd year: Clones of the selected plants are grown in progeny row fashion. Generally due to non-availability of propagule in large amount and to assess a huge number of clones at a time, progeny rows are grown unreplicated. Breeders visually evaluate the clones for desirable traits. Undesirable clones are eliminated and rest are selected for preliminary yield trial.

3rd year: Replicated preliminary yield trial of selected lines is conducted as at this stage the progeny of selected clones provide sufficient propagules. One or few control varieties are also included in this trial for comparison. According to the breeding objective, these clones can be screened separately in disease nursery for prevalent disease and pest. Susceptible clones are rejected. Promising clones are further advanced for multilocation trial.

4th to 6th year: the superior clones are evaluated in replicated yield trials at several locations. The clones are strictly evaluated for yield, quality traits, disease-pest resistance, adaptability traits and compared with the locally adapted check varieties. The best performing clone is selected and multiplied for release as a new variety.

In first two to three generations of clonal selections elimination of inferior plants are done, rather than stringent selection. Selection of individuals for quantitative traits from a population, where individuals of a clone are grown without replication, can yield misleading results. So selections of desirable types are practiced in subsequent clonal generations where replicated trials are conducted.The procedure is given in Fig 6.11.

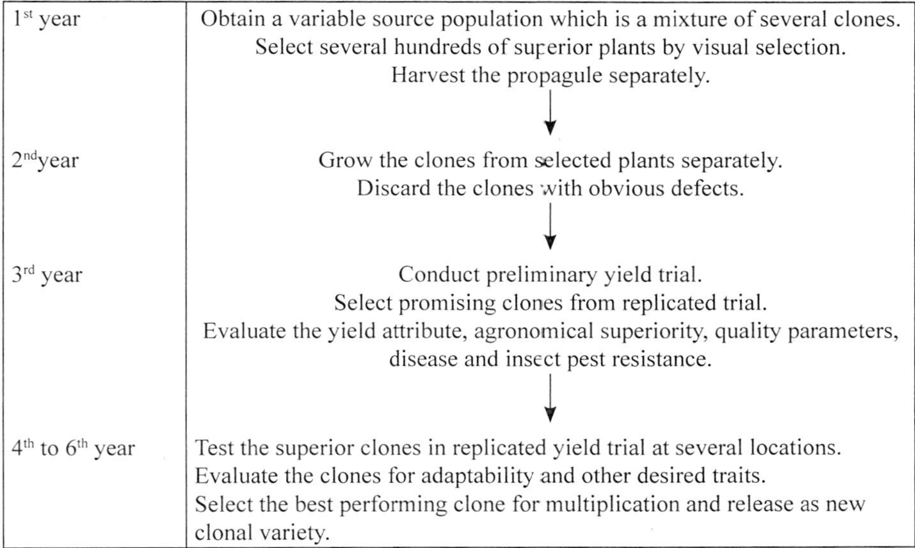

1st year	Obtain a variable source population which is a mixture of several clones. Select several hundreds of superior plants by visual selection. Harvest the propagule separately. ↓
2ndyear	Grow the clones from selected plants separately. Discard the clones with obvious defects. ↓
3rd year	Conduct preliminary yield trial. Select promising clones from replicated trial. Evaluate the yield attribute, agronomical superiority, quality parameters, disease and insect pest resistance. ↓
4th to 6th year	Test the superior clones in replicated yield trial at several locations. Evaluate the clones for adaptability and other desired traits. Select the best performing clone for multiplication and release as new clonal variety.

Fig. 6. 11: A generalized schematic diagram of clonal selection

Hybridization with selection

Hybridization create genetic variability through gene recombination. The population derived from crossing two or more clones is used as base population for selection. The main purpose of this method is to combine desirable characters in an individual from different clones. This is strictly applicable to those clonally propagated species where flower and seed set through sexual reproduction is normal. The detailed procedure is as follows:

Selection procedure

The first step of hybridization programme is selection of parents. The breeding value of F_1 progeny depends on the mean performance of the parents. Genotypes with high general combining ability (GCA) and Specific combining ability (SCA) produce superior F_1 progeny during crossing.

1st year: Clones with high GCA value are crossed to produce F1 seed. The cross may be bi-parental (involve two clones) or polycross (involve more than two clones).

2nd year: F_1 seeds are space planted for better observation of all the plants individually. Unlike the F_1 generation of self-pollinated crops, this is the first segregating generation from where selection process starts. F_1 generation of clonal crops is comparable to the F_2 generation of self-pollinated species. Weak and disease susceptible plants are eliminated. Mild selection intensity is maintained as selection based on single plant performance is not reliable. Large number of plants are selected and their vegetative propagule are harvested separately.

3rd year: Unreplicated progeny rows of selected plants are grown from vegetative propagules. Popularly cultivated clonal variety of that area can be sown for comparison of test lines. Clones with inferior performance and obvious defects are eliminated.

4th year: A replicated yield trial is conducted with all the selected clones and suitable check varieties. An additional screening in disease nursery may also be performed. Few outstanding clones are selected for further evaluation at several locations.

5th to 7th year: The promising clones selected from replicated yield trial are now rigorously tested for yield, quality parameters and resistance at several locations. One or few outstanding clones are selected and gone under varietal release process.

Advantages and limitations of clonal selection

Advantages

1. It is the quickest method to exploit the available genetic variation.
2. Stabilization of genetic constituent of crops can be achieved in a single generation as clones of an individual are true to type. Thus it reduces the time for crop improvement programme.
3. Heterosis can be fixed with minimum efforts and heterozygosity can be maintained as such indefinitely.
4. Clonal cultivars are highly uniform which is very much desirable in commercial cultivation.
5. Sterility is not a bottleneck for crop improvement where seed is not an economically important part.

Limitations

1. Genetic uniformity of clones make them vulnerable to new diseases and pests.
2. Bulky nature of clonal propagules require considerable space during storage.
3. Propagules lose viability and vigour rapidly during long term storage as they are succulent in nature. They demand sophisticated conservation technique.
4. Reduced or no flowering, problems in seed formation and sterility deter the varietal improvement process through hybridization.

Application of clonal selection in spice breeding

Clonal selection is the most important selection method for improving spice crops as just like most of the horticultural crops, spices are also propagated by vegetative means. A large number of popularly cultivated spice crop varieties have been developed through clonal propagation. A few example of such are black pepper varieties Panniyur 6 (selection from Karimunda), Sreekara (selection from Karimunda), Subhakara (selection from Karimunda), Panchami (selection from Aimpiriyam), Pournami (selection from Ottaplackal), IISR Thevam (selection from Thevamundi), PLD-2 (selection from Kottanandan). Cardamom varieties Mudigere 1 (selection from Malabar type), Mudigere 2, ICRI 2 (selection from Mysore type), IISR Vijetha (selection from Malabar type); Ginger varieties Suprabha (selection from Kunduli local), Suruchi (selection from Kunduli local), Himgiri (selection from Himachal collection); Turmeric varieties BSR 1 (selection from Erode local), Krishna (selection from Tekurpeta collection), Sugandham, Roma, Suroam, Ranga (selection from Rajpuri local), Rasmi (selection from Rajpuri local), Pant Peetabh, Suranjana, IISR Kedaram, Kanthi, Sobha, Varna; Cinnamon varieties YCD 1, IISR Nithyashree; Nutmeg variety IISR Viswasree.

6.6 Mutation Breeding

During extensive breeding programme when genetic variability of crop is exhausted, new divergent sources of variation has to be sorted by breeder. In this purpose, beside hybridization, mutation is another avenue to create genetic variability for facilitating crop improvement. Mutation is considered as the ultimate source of any variation. Each cultivated species, in its transformation from wild form to cultivated form during course of evolution, have accumulated a number of beneficial mutations in its genome.

In broad sense, mutation is considered as the sudden heritable change in the genetic material at the gene or chromosome level. Change in hereditary material is associated with deletion, duplication, inversion, translocation of chromosome or change in nucleotide level (point mutation). Mutation can occur naturally at very low rate by cosmic or ultraviolet ray, called spontaneous mutation. Induced mutation is artificially induced (through human intervention) mutation using physical or chemical agent. The agent used to induce mutation is called mutagen and the mutated individual with altered character(s) is mutant.

Mutation breeding refers to the use of induced mutations in development of improved varieties. The situations when the mutation breeding will be the best option to follow are:

1. When there is no natural variability available in the gene pool of the crop species for the desired character to be improved.
2. When a desired trait is so tightly linked with undesirable trait that recombination between them through hybridization is not possible.
3. When variation is to be created in clonally propagated species where sexual reproduction is not normal.
4. When one or two simply inherited traits are to be modified in an otherwise superior cultivar.

Mutation breeding is applicable to all the crop species but effectiveness of the mutation induction depends on the breeding behavior of the species. It is more successful in self-pollinated species than cross-pollinated ones. More pronounced application of mutation has been seen in vegetatively propagated crops where sexual hybridization is difficult.

Classification of mutagens

A large array of physical and chemical mutagenic agents are used in modern crop improvement. Different classes of mutagens and their examples are as follows (Fig 6.12).

Physical mutagen	
Ionizing radiations	X ray, Gamma ray (γ ray), Alpha ray (α particle), Beta ray (β particle), Fast neutron
Nonionizing radiations	Ultraviolet ray (UV ray)
Chemical mutagen	
Alkylating agents	Ethylmethanesulphonate (EMS), Methylmethanesulphonate (MMS), Ethylethanesulphonate (EES), Nitrogen mustard, Mustard gas
Acridine dyes	Ethidium bromide, Acridine orange, Acriflavin, Proflavin
Base analogues	5 Bromouracil, 2 Aminopurine
Deaminating agents	Nitrous acid
Azides	Sodium azide

Fig. 6.12: Classification of mutagenic agents

Among different mutagens, ionizing radiations (X ray, Gamma ray) are most frequently used in breeding programme due to their high penetration power, reproducibility and above all easy to application. They mainly cause chromosomal aberrations through breaking parts of chromosome. Whereas, alkylating agents are preferred among chemical mutagens. Unlike physical agents, chemical mutagens generally bring changes in genome through point mutations. Major drawbacks associated with using chemical mutagens are their poor reproducibility, uneven penetration, presence of toxic residuals.

Optimization of dose for the mutagen with respect to concentration and length of exposure is very important. It varies with plant species, plant parts to be treated, physical environment etc. A mutagen dose which restricts the survival to 50% (LD_{50}) or growth to 50% (GR_{50}) is considered the optimum.

Choice of plant material to be mutagenized

Generally all plant parts can be treated through one or the other mutagen. But treating the seed is more common in practice. The seed soaked in water is active in DNA synthesis, which is ideal for induction of mutation. In vegetatively propagated crops, treatments are given to sucker, bulb, tuber, rhizome, corm or any other parts used for propagation.

The success of attaining desired level of mutation depends on a number of internal and external factors, such as temperature, pH, oxygen level of the environment, moisture content in the target tissue etc.

Procedure

1st year/M1 generation: The parent genotype to be induced mutation is chosen which is an advanced line or adapted new cultivar with specific limitations. The true to type seeds (M0) are treated with mutagen of optimum dose. Treated seeds (M1) are planted immediately after treatment. Untreated control plant population should be grown for comparison of the mutagen treatment effect in germination, lethality, growth, M1 injury and sterility. The number of M1 plants to be grown are adjusted considering lethality % of the treatment and expected M2 population size. M1 population can be close planted to restrict the growth of secondary branches as maximum genetic variability due to mutation is expected in primary branches. In other way close planting helps to accommodate maximum M1 plants which increase the chance of getting desirable variants in M2 generation. Selfing of all M1 plants is to be ensured by planting in isolation or bagging the inflorescence prior to anthesis. Otherwise, variation created due to out-crossing and mutation cannot be separated in segregating generation. Records on germination, seedling growth, chimera induction, survival to maturity, sterility are taken. Chimera is the state where an individual is consist of cells of more than one distinct genotypes. The most common visible form of chimera in M1 plant is leaf chlorophyll variegation. Each M1 plant is harvested separately.

Generally mutation is recessive in nature. So it is not expressed in heterozygous M1 generation. If mutation is dominant, then it is expressed in M1 generation itself and selection for mutant is practiced in this generation.

2nd year/M2 generation: individual plant progeny rows of each M1 plant are grown. The parent genotype is grown at regular interval for easy identification of mutants. Selection of mutant traits are practiced as recessive mutations start to appear in homozygous from M2. In cross-pollinated crops, mutant allele is still in heterozygous in M2 generation. So another year of selfing is done to produce homozygous M3 progenies. The size of M2 population should be as large as possible to harbor maximum mutant types in this segregating generation. Individually plants are selected by visual identification of mutant phenotype (disease susceptibility, earliness, fruit character) or by physical (seed shape, size) or biochemical (alkaloid, protein, nutrient content) methods. Desirable variants are harvested individually.

3rd year/M3 generation: progeny rows from the selected M2 plants are grown. Progeny test in M3 generation is essential for testing the heritability of putative mutant phenotype selected in M2. It is also necessary to stabilize the performance of potential mutants. Selection for quantitative traits are practiced in M3. Selection for potential homozygous mutants in cross-pollinated crops also starts here. Progeny rows with undesirable mutants are eliminated. Desirable mutant M3 rows are harvested in bulk if sufficient homogeneity is achieved.

4th year/M4 generation: a preliminary yield trial of potentially superior mutant line is conducted with check varieties. Promising lines are advanced to multilocation trials.

5th-7th year/M5-M7 generation: promising mutant lines are tested in replicated multilocation yield trial. Outperformer is released as new variety.

Potential variants for other than target trait(s) of interest of that particular mutation breeding scheme are maintained separately and can be used in hybridization programmes.

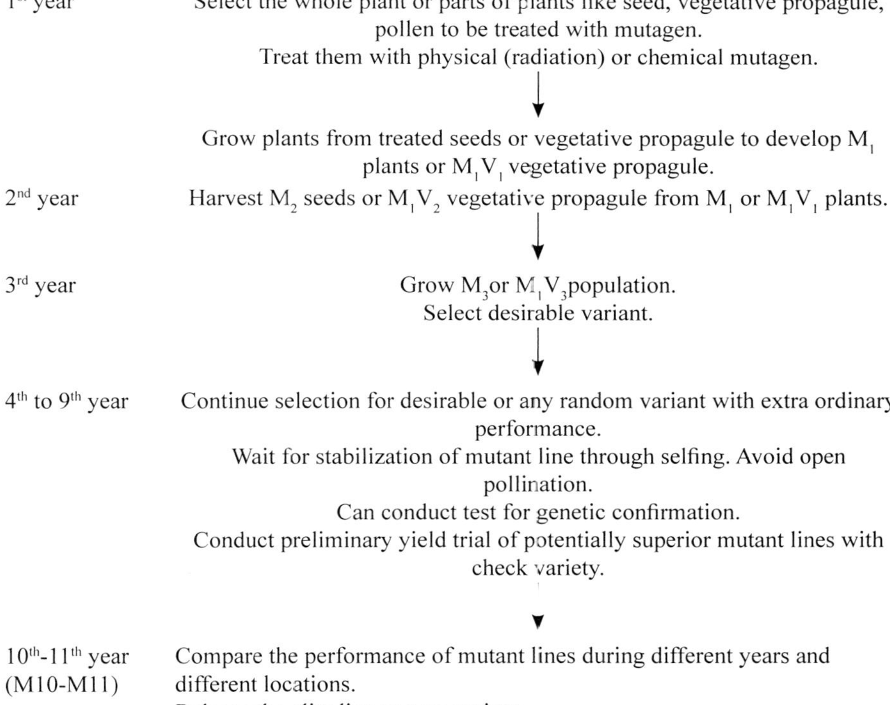

Fig 6.13: A generalized schematic diagram of mutation breeding in crop improvement.

Application of mutation breeding in spice crops

Mutation breeding is a special type of breeding method can be applied in any sexually or asexually propagated crops. Spice crops are not exception where crop improvement through mutation breeding has already been started. Suravi is an induced mutant variety of ginger developed from Rudrapur local variety. Co 1 is an X-ray irradiated mutant turmeric from Erode local variety. Coriander variety RCr 684 by gamma irradiation of RCr 20, cumin variety RZ

223, fenugreek varieties RMt 303 and RMt 305 (from RMt 1) are some other achievements.

6.7 Ploidy Breeding

The somatic chromosome number of any species, whether diploid or polyploidy is designated as 2n, and the chromosome number of gamets is denoted as n. An individual carrying the gametic chromosome number, n, is called haploid. A monoploid has the basic chromosome number, x.; one x constitutes a genome or chromosome complement. A diploid species has two, a triploid has 3 and a tetraploid has 4 genomes and so on.Thus

Since the triploids are generally sterile and they do not produce seed but may set fruits. Triploid seeds are produced by crossing 2n x4n=3n. Triploid seeds are given to farmers along with pollinator (2n). So, production of triploid seeds calls for maintaining diploids and triploid line for producing triploid seed every year.

Tetraploids produced from their diploid progenitors are generally sterile / lessfertile but may be more vigorous vegetatively.

Aneuploids

It means not true polyploids. Aneu mean 'false', ploid =levels (folds) such asn + 1; 2n + 2etc.

Autopolyploidy

Monoploidy, triploidy, tetraploidy and higher levels of ploidy are included in autopolyploidy. Production of monoploids and haploids is entirely different from that of other ploidy levels. These two do not constitute polyploidy.

Other autopolyploidy are produced directly or indirectly through chropmosome doubling. This doubling may be spontaneous in somatic tissues and unreduced gamets are produced. Again, this doubling may be through physical agents viz., heat or cold treatments, centrifugation, x-ray or gamma-ray irradiation, etc. Polyploidy is a common feature of the cells during in vitro culture.

Colchicine treatment is the most efficient and widely used treatment for chromosome doubling. Colchicine is actually a poisonous chemical isolated from seeds and bulbs of *Colchicum autumnale* (autumn crocus). Pure colchicines is $C_{22}H_{25}O_{6}N$. It blocks spindle formation and thus inhibits the movement of sister chromatids to the opposite poles.As a result the chromosome number of the cell is doubled.Since colchicines affects only dividing cells, it should be applied when the tissues are actively dividing.It is important to use freshly prepared aqueous solutions of colchicines through seed treatment, seedling treatment (only young shoots are to be treated, not the roots), growing shoots apice trteatment (by brush or with a dropper), etc Other chemical agents having polyploidizing effect are namely acenaphthene, 8-hydroxyquinoline and nitrous oxide.

Allopolyploidy

Allopolyploids have genomes from two or more species. It basically aims at creating new species. This can be achieved by doubling the chromosome number of distant hybrids with the help of colchicines or some other agents. It is estimated that about one-third of the Angiosperms are polyploids, and by far the vast majority of them are allopolyploids.

Application of ploidy breeding in spice crops

Polyploidy is useful in cases when the crossing between two species is not possible because of differences in ploidy level as a bridge for gene transferring between them.Autotetraploid plants were induced in Trigonella foenum-graecum by means of colchicine.The induction of autotetraploids in Trigonella foenum-graecum unfortunately apparently resulted in marked sterility which reduced yield of seed. The high percentage of pollen sterility accompanied with very poor setting of seed has been attributed, mainly, to the abnormalities recorded at anaphase I and II.

Cardamom being perennial, development of homozygous line should take a long time considering the number of selfing cycles and generations involved. Anther and microspore culture may provide a shortcut for development of haploids and dihaploids and reduce the time considerably.

Development of polyploidy in garlic (*Allium sativum L.*) was reported in china. Polyploidization is one of the important tools as alternative approaches for genetic improvement of plant species.In adverse environment, the physiological as well as biochemical flexibility enable polyploid plants to possess stronger tolerance and viability than the diploid progenitors or relative species, especially for the vegetatively propagated or perennial plants. Polyploidization in garlic is induced by colchicine and the chromosomes doubling is achieved by different concentration of colchicine.

Vanilla planifolia display important phenotypic variation, but little genetic diversity is demonstrated by AFLP and SSR markers. An autopolyploidy in V. planifolia contributing to the phenotypic variation has been observed.

7

Next Generation Plant Breeding

7.1 Plant Tissue Culture

Spices have great socio-economic impact in moden food habit of human but loss of resources day by day due to various reasons (industrialization, high population pressure) make the access of these valuable spices limited to people. Low production is general phenomena of common spices due to lack of high yielding varieties. Hence there is a need to improve the propagation method to meet the gap between demand and supply. Conventional vegatative propagation method is common approach for propagation but low multiplication rate, seasonal dependency and susceptible to abiotic stresses and diseases are major problems to achieve the optimum productivity. To overcome these drawbacks scientists have chosen the plant tissue culture technique (a modern approach of biotechnology which deals with the propagation the valuable/endangered plants under laboratory condition). Details about this technique and its role on spices are given below.

Plant Tissue Culture: Basic Components and Objectives

The technique of growing of a whole plant cell, tissue or organ under an aseptic condition in suitable synthetic medium is termed as plant tissue culture (Thorpe, 2007). De-differentiation and re-differentiation are two essential fundamental processes which help in the physio-morphological development of the plant.

Components of plant tissue culture: Morphological developments in the plant during plant tissue culture are controlled by three essential components namely - explant, culture medium and cultural environment. The small plant part detached from the mother plant is used to initiate the culture under *in vitro* condition is termed as explant. The synthetic supplement which supports the growth and development of the plant from an explant under *in vitro* condition is called medium (*plural: media*). On principle, a synthetic medium is composed of various components similar to essential elements of soil which allows the plant to grow at their natural condition. So composition of the medium is the determining factor for proper the growth and development of explants under *in* *vitro* condition. There are no specific media available to carry out the culture of all types of plants under *in vitro* condition. Selection of medium depends upon the type of culture to be initiated, objective of culture and types of explant. For example, Nitsch and Nitsch media (1969) is applicable for haploid plant production while MS media (Murashige and Skoog, 1962) is widely used for micropropagation of various plant species. Environmental parameters such as light, temperature and relative humidity that affect the growth and development of plant at *in vitro* condition are considered as cultural environment.

Fundamental objectives of plant tissue culture: Though fundamental concept on plant tissue culture was first postulated by Gottlieb Haberlandt in 1902 but justified successfully by Gautheret (1934) for the first time during *in vitro* culture of cambial tissue of *Acer pseudoplatanus* (Thorpe, 2007). Since plant cells have capacity to regenerate into whole plant, based on this "totipotent" property scientists developed protocols for *in vitro* propagation of several plants using the "explants".

Irrespective of plant species, basic application of plant tissue culture broadly covers several aspects of plant breeding involves- i) Production of genetically identical plant (through the process of micropropagtion or somatic embryogenesis method); ii) Plant genome transformation (using the technique of genetic engineering; iii) Creation of novel varieties through several approaches like variants against different stress (based on somaclonal variation; transgenic breeding), production of somatic hybrid (based on protoplast fusion), and homozygous line (based on haploid culture); iv) Production and enhancement of secondary metabolites under *in vitro* condition (Debnarh *et al*., 2006; Altpeter *et al*., 2016; Espinosa-Leal *et al*., 2018) and v) Germplasm conservation (Sengar *et al*., 2010; Hussain *et al*., 2012) which is generally done at ultra low temperature (cryopreservation).

Plant Tissue Culture in Spices

Micropropagation*:* Though conventional propagation methods are common approch for plant propagation but are unable to achieve the optium productivity due to presence of several external biotic and abiotic factors. To overcome these problems, "micropropagation" has been invented by scientists. Commercial impact of micropropagtion came to reality for the first time when Morel developed micropropagation technique for Orchid (1960). It is a modified version of vegetative propagation method which is performed aseptically under *in vitro* condition at controlled environment using an artificial growth supporting medium for an explant. Hence this technique is beneficial for those type of plants which have either low multiplication rate or for which the seed based propagation is difficult

(Jeong *et al.*, 1995; Xiao *et al.*, 2011). This technique is advatageous over the conventional vegetative propagation method due to higher multiplication rate and ability to produce the infection less plants (virus free) (Gantait *et al.*, 2011). In addition, it permits season independent supply of genetically identical plants from a single explant.

Micropropagation method is multistage procedure, comprises of three distinct phases i.e. Stage I (culture establishment stage); Stage II (multiplication stage) and Stage III (rooting stage and preparation of transferring the plantlets to outside from *in vitro* condition) (Murashige, 1974). Being part of this process these stages have also decoded points where the environmental or media changes are needed (Torres, 1989). In addition to these stages, extra two different stages have been added by Debgarh and Maene (1981) as Stage 0 which involves the preparation of the stock or mother plant to supply infection free quality explants during experiment and Stage IV which permits transfer of the *in vitro* plantlets from laboratory to field condition.

Success of micropropagation depends upon several factors such as nature of explants (genetic makeup, physiological condition, age of mother plant etc.), cultural environment as well as medium components. Though micropropagation protocol has been standardized for spices using different type of explants but most of the cases it is specific to spice itself i.e. buds from rhizome has been reported as explants for micropropagation of *Zingiber officinale* Rosc. (Balachandran *et al.*, 1990; Ilahi and Jabeen, 1992; Rout *et al.*, 2001; Kambaska and Santilata, 2009); red ginger (*Zingiber montanum* Koenig) (Hamirah *et al.*, 2010); turmeric (Balachandran *et al.*, 1990; Salvi *et al.*, 2002; Nasirujjaman *et al.*, 2005; Singh *et al.*, 2011; Nayak *et al.*, 2011) while node has been reported for mentha (Ghanti *et al.*, 2004; Vasile *et al.*, 2011; Mehta *et al.*, 2012; Gomes *et al.*, 2015). In addition to these explants, other different explants have also been reported for several spices e.g. shoot tip for *Elettaria cardamomum* Maton (Tyagi *et al.*, 2009); *Zingiber officinale* Rosc. (Ibrahim *et al.*, 2015); leaf, node and apical meristem for *Piper nigrum* L. (Hussain *et al.*, 2011).

Among three major components of tissue culture mentioned earlier, medium can be manipulated. Most of the *in vitro* propagation methods for spices are based on MS (Murashige and Skoog, 1962) medium. Murashige and Skoog (MS) medium was reported as more responsive than Nitsch and Nitsch (NN) and Gamborg (B5) media for *in vitro* propagation of Oregano (*Origanum syriacum* L.) (Abdallah *et al.*, 2017); Similar response was reported by Bejoy *et al.*, 2012. They found maximum multiplication under MS media with respect to other media (Nitsch, Woody Plant medium (WPM), Gamborg (B5) and White media) during micropropagation of Indian ginger. Though contradictory report also was reported by Musallam *et al.*, 2011 for *in vitro* propagation of caper (*Capparis spinosa* L.) where they found WPM was better than MS medium during culture establishment of caper.

Essential elements of medium provide basic nutrition to explants while proper morphogenesis is carried out in presence of optimum level of plant growth regulators (PGRs). Plant growth regulators are synthetic compounds which are effective at low concentrations to promote either inhibition or modification in the plant morphogenesis. Though basic function of PGRs and plant hormone is same but difference lies on their origin i.e. plant hormone is natural substances while the PGRs are synthetic in nature (Davies, 2010). First significant impact of plant growth regulators (PGRs) in plant tissue culture was reported by Skoog and Miller (1957) where they studied the role of auxin and cytokinin in plant morphogenesis of tobacco.

Their hypothesis suggested high concentration of cytokinin over auxin induced *in vitro* shooting in explant while reciprocal ratio (auxin > cytokinin) and equal ratio of auxin and cytokinin in explant promoted *in vitro* rooting and an intermediate "callus" respectively. Based on these findings, scientists developed micropropagation protocols for several spices using different types of explants e.g. *Satureja punctata* using shoot tip from *in vitro* germinated seeds (Teshome *et al.*, 2016); *Zingiber officinale* Rosc.' variety Tambunan' using rhizome bud (David *et al.*, 2016); *Piper chaba* Hunt. using node (Rani and Dantu, 2016); capsicum (*Capsicum annuum* L.) using cotyledon (Hegde *et al.*, 2017a); *Mentha piperita* L. using shoot tip and node (Islam *et al.*, 2017); *Piper Longum* L. using node (Fonseka and Wickramaarachchi, 2018); Tunisian local garlic (*Allium sativum* L.) using shoot tip (Ayed *et al.*, 2018); pepper mint (*Mentha piperita*) using node (Sharma *et al.*, 2019); Lota Bhot variety of (*Capsicum chinense* Jacq.) using seed and node (Bora *et al.*, 2019); black galingale (*Curcuma Aeruginosa* Roxb.) using shoots from rhizome (Khumaida *et al.*, 2019); *Piper sarmentosum* using node as explants (Stephin *et al.*, 2020). Indirect approach of micropropagation (plant regeneration through callus) using leaf have been reported for chilli (*Capsicum chinense* Jacq.) (Bora *et al.*, 2018); *Mentha piperita* (L.) (Islam and Alam, 2018); *Zanthoxylum armatum* DC. (Purohit *et al.*, 2019); Disease free plant productions through plant tissue culture method have been reported for several spices. Bhattacharya and Sen (2006) developed disease free *Zingiber officinale* Rosc plant from sprouted bud of rhizome and found no infection in rhizome (rotting) of micropropagated plant when kept it for 6 month in sand while 61% of the rhizomes of mother-derived plants rotted during storage. Similar findings shared by Sharma and Singh (1997) where they found no rotting in rhizomes micropropagated ginger in compare to 54.5% rotted rhizome from mother plant during storage.

Somatic Embryogenesis: Somatic embryogenesis is a developmental pathway in which somatic cells convert into embryogenic cells. These embryogenic cells pass through a series of morphological and biochemical changes under certain conditions, form somatic embryos and may be converted into whole plants (Schmidt *et al.*, 1997; Komamine *et al.*, 2005; Yang and Zhang, 2010). Though somatic embryogenesis follows same developmental pattern as zygotic embryos (globular, heart , torpedo, and cotyledonary stages for dicots, or globular, scutellar, and coleoptilar stages for monocots) (Gray *et al.*, 1995; Toonen and de Vries, 1996; Jiménez, 2001) but difference lies on their origin during zygotic embryogenesis, the product of fertilization (zygote) follows embryogenic developmental pathways while in somatic embryogenesis, somatic cell has to achieve the ability to respond to certain signals/stimulants and then only embryogenesis happens (Pasternak *et al.*, 2002). Most unique advantage of somatic embryogenesis over micropropagation permits production of plants with pre-build shoot and root system (Shrikhande *et al.*, 1993). As a result researchers can avoid "the rooting stage" of micropropagation which actually makes the procedure simple, cost effective and less time taking.In addition, encapsulation of somatic embryo acts as "artificial seed" which allows the exchange of germplasms from one place to another easily. Due to these significant advantages, somatic embryogenesis has gained attention by the plant biotechnologist and also is also considered as substitute of micropropagation.

This developmental pathway of somatic embryogenesis can be divided into two phases – at initial phase cells are induced to form embryo (induction of somatic embryo) followed by formation of competent cells to form somatic embryo during later phase (expression) of embryogenesis (Jimenez, 2001; Namasivayam, 2007). Like microproapagtion, induction and development of somatic embryogenesis from explants are based on several conditions:

According to Emons (1994), stages and basic conditions to be maintained for development of somatic embryogenesis from explants are given below

i. ***Somatic embryo induction***: Dedifferentiation of cells in the explants at high auxin rich medium. Most widely used auxin for induction of somatic embryo is 2, 4-D.

ii. ***Development of somatic embryos***: Initiation of embryogenic developmental pathway in competent cell starts from a globular stage in presence of medium having very less/no auxin.

iii. ***Somatic embryo maturation***: Embryo maturation with well-developed embryo storage organs can be achieved by the application of a maturation medium with high sucrose concentration.

iv. ***Prevention of germination*** before embryo maturation using abscisic acid.

Presence of auxin is prerequisite at initial stage of somatic embryogenesis to establish an auxin gradient to initiate bilateral symmetry in cell during embryogenesis (Schiavone and Cooke, 1987; Liu *et al.*, 1993; Fischer and Neuhaus, 1996; Jime′nez, 2005). Presence of auxin for long time accumulates endogenous ethylene which indirectly enhances cellulase and pectinase activity. Components of cell wall are susceptible to both of these enzymes hence prolong treatment of auxin prevents proper development of embryo (Dantu and Tomar, 2010). Though there are different types of explants (seedlings petioles, leaves, roots, shoot meristems, seeds, cotyledons, immature zygotic embryo and zygotic embryos) are available for somatic embryogenesis but most frequently used explants is immature zygotic embryo (Gaj, 2004). Somatic embryo in plant cell under *in vitro* condition may be achieved through either direct approach (without callus) or indirect approach (callus present as an intermediate) (Hossein *et al.*, 2006). Both of direct and indirect approach of somatic embryogenesis has been studied by Shajahan *et al.* (2016) for mango ginger (*Curcuma amada* Roxb.). Plant regeneration through somatic embryogenesis method has been reported for several spices including *Crocus sativus* L. (Sevindik and Mendi, 2016); six varieties of black pepper (*Piper nigrum* L.) (Sasi and Bhat, 2016); *Capsicum baccatum* (Venkataiah *et al.*, 2016); Co-1 and Rajendra Swathi (RS) varieties of *Coriandrum sativum* (Ali *et al.*, 2017); *Kelussia odorotissima* Mozaff. (Ebrahimi *et al.*, 2018); turmeric (Jie *et al.*, 2019); *Allium sativum* L. (Malik *et al.*, 2020). Formation of synthetic seeds in somatic embryo has been reported for *Mondia whitei* (Baskaran *et al.*, 2015); *Coriandrum sativum* L. varieties (Rajendra Swathi 'RS' and Co-1) (Ali *et al.*, 2018a). Besides true to the type plant production, somatic embryogenesis has been reported for secondary metabolites production. Firoozi *et al.* (2018) reported role of PGRs in secondary metabolites production (picrocrocin, safranal, crocin) during somatic embryogenesis of saffron.

Protoplast Fusion viz a viz Somatic Hybridization: Conventional hybridization method is common approch in crop improvement but does not applicable to sexually incompatible plant speices. Since plant cells have the totipotent property, then protoplast may be used as source material to perform the same function. Besides protoplast can uptake the foreign DNA through their membrane under certain conditions while fusion of protoplast from different source results hybrids. So sexual incompatibility between distantly related species may be overcome through this technique (Tomar and Dutta, 2010). From these above metioned informations, we can divide the role of protoplast either as explant (for protoplast culture) or material for fusion at cellular level (protoplast fusion to create somatic hybrids).

i. **Protoplast culture :** A new era of plant biotechnology began when first successfully isolation of protoplast was reported by Cocking *et al.* (1960) for tomato. Though protoplast isolation may be carried out through either mechanical or enzymatic method but enzymatic approach is best to get viable and good quality protoplast. Leaf has considered as most suitable explant for protoplast isolation due to presence of relatively thin walls of mesophyll cells which are easily degraded by enzymatic treatment. In most of the cases, cells are treated with mixtures of both pectinase and cellulase enzymes where pectinase acts on pectin (component of middle lamella of cell wall) and cellulase acts on cellulose of cell wall. As consequences, cell becomes fragile and susceptible to osmotic damage. To overcome this problem, osmoticum such as mannitol or sorbitol is commonly added to during experiment on protoplast isolation as well as protoplast fusion and somatic hybridization and is removed from the media after formation of cell wall. Isolated protoplast can be used as explants to obtain plant (first reported by Takebe *et al.*, 1971 for tobacco) while fusion of viable protoplast results somatic hybrids (first reported by Carlson *et al.*, 1972 for hybrids of *Nicotiana glauca* and *N. langsdorffii,*). Conversion of protoplast into whole plant depends on several factors such physical condition as well as status of cell enzyme to be used, treatment duration and concentrations of enzyme, media components etc (Davey *et al.*, 2005). Reports on protoplast culture for spices are very less. Use of macerozyme R-10 and cellulase Onozuka R-10 have been reported to isolate protoplast followed by protoplast culture to regenerate whole plant for several plant spices including pepper mint (*Mentha piperita* L.) (Sato *et al.*, 1993); ginger (*Zingiber officinale* Rosc.) (Guo *et al.*, 2007). Application of cellulase, pectinase, macerozyme and driselase were reported to isolate protoplast from embryogenic suspension of Indian local *Coriandrum sativum* L. followed by protoplast culture and regeneration of plant (Mujib *et al.*, 2014). Macerozyme R-10 and cellulase Onozuka R-10 was also reported by Geetha *et al.* (2000) to isolate the protoplast from leaf for protoplast culture of cardamom (*Elettaria cardamomum* Maton.) and ginger (*Zingiber officinale* Rose.).

ii. **Protoplast Fusion and Somatic Hybrids:** Since the cell membranes have negative charge hence fusion of two protoplast cells is very difficult. To overcome this problem, fusions of protoplasts from different sources are mostly carried out through either chemical agent (fusogen) or physical method (electrofusion). Though rapid and highly productive fusion can be possible through electrofusion method within short time (Bates *et al.*, 1983) but most of the reports for spices are based on chemical mediated fusion. Polyethylene Glycol (PEG) mediated effective protoplast fusion has been reported by Guan *et al.*(2010) for ginger (highest binary fusion rate found for the fusion of ginger ['Lushan Zhangliang jiang' + 'Chenggu Huang Jiang' (LZ + CH)] but the fusion of ginger ['Chenggu Huang Jiang' + 'Sichuan Zhugen Jiang' (CH + SZ)] regenerated the plantlets). PEG mediated protoplast fusion have also been reported for fusion between tansy (*Tanacetum vulgare* L.) and pyrethrum (*Tanacetum cinerariifolium* (Trevir.) Schultz-Bip.) (Keskitalo *et al.*, 1999); *Vanilla planifolia* and *V. andamanica* (Divakaran *et al.*, 2008). That may be due to of low toxicity of PEG to most of the cell types as well as ability to produce the heterokaryon at high frequency (Kao and Michayluk, 1974; Durieu and Ochatt, 2000; Guan *et al.*, 2010). Nature of protoplast fusion is depending upon the concentration of polyethylene glycol and treatment duration. The major advantage of protoplast fusion is the production of cybrids i.e. cell consists of the single nucleus and the cytoplasmic genome of one parent while another contributes the only cytoplasmic genome (Chiang *et al.*, 1993) but till now no report is available from spices on this aspect. of protoplast fusion.

Haploid culture: Plant breeding progamme deals with crop improvement and crop productivity by integration of new trait to plant as well as the genetic gain to increase the yield. Development of pure line from segregating populations is a conventional way to track the genes for a trait of interest. But creation of pure line is time consuming through traditional method. Basic method of pure line development can be done through either selfing of F1 population upto 7-8 generation (generates Recombinant Inbred Lines or RIL) or backcrossing up to 6-7 generation (generates Near Isogenic Lines or NILs). In compare to these conventional methods, production of homozygous line or double haploid can be achieved through tissue culture technique within very short time. Development of haploid line provides breeders an excellent opportunity to study the genes at the hemizygous condition and also help to identify the variations (mutations) (Morrison *et al.*, 1991). Three common practices that are commonly used to produce of haploids and homozygous line or double haploids under *in vitro* conditions are - androgenesis (involves microspore culture; anther culture); gynogenesis (involves ovary or ovule culture but labour intensive) and chromosome elimination (Yan *et al.*, 2017). Microspore culture is considered as better option than anther culture because of free from negative impact of anther wall (lack of anther wall allows more uptake of nutrients which helps in proper development of microspore) and less time consuming than anther culture (Ferrie and Caswell, 2011) but to perform this culture needs better skilled persons and sophisticated techniques (Germanà, 2011). Through gynogenesis and androgenesis, genome wide homozygosity can be achieved in a single generation programs which directly reduces the breeding time (Dunwell, 2010; Germanà, 2011; Portemer *et al*., 2015). The idea of production of haploid plant through tissue culture technique came to reality when Guha and Maheswari (1964) reported production of haploid plant using anther (containing immature pollens) as explants for *Datura innoxia*. Double haploid method has been considered as effective tool to fixed the genetic variation within short time span and permit production of completely homozygous from the heterogeneous source. Basic criteria to be considered for this purpose lies on production of large numbers of haploid plants and efficient chromosome doubling (Ślusarkiewicz-Jarzina *et al*., 2017). Among the several factors, stage of microspore is most important. For haploid production, microspore between late uninucleate to the mid binucleate stage is are commonly used but most effective stage is uni-nucleate stage prior to starch accumulation (Acharya and Ramji, 1977) . The anther having pollens at late uni-nucleate stage formed maximum callus followed by anthers having microspore at early uni-nucleate stage during *in vitro* andrgeneis of capsicum (*Capsicum annuum* L.) (Hegde *et al*., 2017b). Smýkalová *et al*. (2009) suggested anthers between mid-uninucleate to early binucleated cells was best for *in vitro* response in caraway (*Carum carvi* L.). Though anther having more than 50% late unicellular microspore was reported as effective explants to produce haploid plant for hot pepper (Supena *et al*., 2006). Most of the reports related to androgenesis are available for pepper. Based on agro-morphological characteristics with respect to original cultivar Shrestha *et al*. (2011) found anther derived SP94, SP117, SP148, SP171 and SP174 as superior diploid lines of sweet pepper. *In vitro* androgenesis was studied by Ari *et al*. (2016) for 48 genotypes of *Capsicum annuum* L. and found shed microspore culture as effective method for production of double haploids. Use of colchicines for diploidization has been reported by Olszewska *et al*. (2015) where they used four different hybrid i.e. (905 × 'Sono') F1, (905 × 'Mino') F1 , (405 × 'Luba') F1 , (405 × 'Sono') F1 of pepper and achieved an effective diploidization at the level of 17-27%. Production of haploid line of black cumin through ovule culture has been reported by El-Mahrouk *et al*. (2018). Production of double haploid *Carum carvi* L. through androgenesis was reported by Smýkalová *et al*., 2009.

***In vitro* conservation of germplasm:** Natural resources of plants are reducing day by day due to uncontrollable human activities. Hence conservation of elite plant germplasms is utmost step to keep them alive in nature. Besides keeping balance in nature, germplasm conservation allows breeders to use the conserved germplasm for future purose. Though conservation may be done through *in situ* method but lack adequate space is major barrier to conserve it properly (Engelmann, 1991). Hence *ex situ* methods is another approach to conserve them out of the natural habitat. Among the various *ex situ* conservation method, *in vitro* conservation method is most widely used to conserve the germplasm.

In vitro conservation may be done either for short term or long term depends upon storage duration. Long term conservation is carried out at ultra low temperature (-196°C) in presence of liquid nitrogen (cryopreservation). Cryopreservation provides safe and cost-efficient long-term conservation of various types of plants (Engelmann, 2011). Cryopreservation is attractive because of several advantages such as cost effective, less space required (Matsumoto, 2001). Basic principle. Basic principle behind this conservation deals with the metabolic activity of cells and ceased the cell divisions at ultra low temperature (Cruz-Cruz *et al.*, 2013). Cryopreservation may be carried out through either freeze induced dehydration or vitrification (Engelmann, 2000; Rao, 2004). But freeze induced dehydration leads to cryoinjury because of formation of ice as well as increase the intracellular solute concentrations within cell at ultra-low temperature. To counter this problem, Vitrification has been developed which allows the transition of water directly from the liquid phase into an amorphous phase or glass, whilst avoiding the formation of crystalline ice (Fahy *et al.*, 1984). Encapsulation to explants provides safety from mechanical injury during cryopreservation (Niino and Sakai, 1992; Suzuki *et al.*, 2005). Synthetic seed due to its small size considered as ideal material for germplasm staorage. Besides, synthetic seed based approach for germplasm conservation is cost effective and permits transfer of germplasm one place to another (Zych *et al.*, 2005). Cryopreservation has been reported for several spices based on synthetic seed based approach. Cryopreservation through droplet-vitrification protocol was found better than encapsulation-dehydration for conservation of mint (Martín *et al.*, 2015). Use of encapsulation dehydration, encapsulation vitrification, and vitrification for cryopreservation of *in vitro* grown shoots of ginger was reported by Yamuna *et al.* (2007).

Secondary metabolites production: Besides plant production, plant tissue culture technique is also deployed to produce the secondary metabolite. Secondary metabolites have no direct role in growth but provide the defence mechanism and the product has economic and significant importance. Accumulation of metabolites often occurs in plants subjected to stresses including various elicitors or signal molecules (Akula and Ravishankar, 2011). Producing secondary metabolites through cell culture is much more advantageous over *in vivo* because of its simplicity, permits more production under control condition than *in vivo*, free from environmental stresses and less susceptible to contamination (Karuppusamy, 2009). For this purpose, cell suspension culture system has been adopted as a continuous, reliable source to produce the secondary metabolites (Vanisree *et al.*, 2004). This method has become popular because of its rapid growth cycles and generate large amounts of cells for quantitative or qualitative analysis of growth responses and metabolism of novel chemicals (Mulabagal and Tsay, 2004; Owis *et al.*, 2019). Inspite of these advantages, major constraints are instability of cell line, poor growth and low yield (Ravishankar and Venkataraman, 1993; Rao and Ravishankar, 2002). Cell suspension culture performed at liquid medium using callus as source material and incubated under shaking condition. Shaking provides uniform distribution of media as well as converts the clumps of cell to individual cell unit. Till now most the reports related to secondary production are based on using elicitors. Elicitors are the compounds which trigger any type of physiological abnormality in the plant system.. These elicitors are categorized as either exogenous (produced by the pathogen) or endogenous (produced within plant system due to the pathogenic effect) elicitors (Sharma *et al.*, 2011). Use of elicitor to culture depends on several factors such as elicitor concentration, growth stage of the culture at the time of elicitor addition and exposure time with the elicitor (Bourgaud *et al.*, 2001; Zhao *et al.*, 2005; Namedo, 2007;Sharma *et al.*, 2015). Use of elicitors (50 g/L sucrose and 50 or 100 mg/L proline) in MS media have been reported to induce the secondary metabolites thymol in callus of *Origanum vulgare* L while NaCl 200 mg/L had negative impact on it (reduced 50.56% thymol production as compared to control) (Al-Jibouri *et al.*, 2012). Role of PGRs such as thidiazuron (TDZ) have been studied as an elicitor during cell suspension culture of Maraş-1 pepper seeds to determine the on total phenolic, total protein, superoxide dismutase (SOD) and catalase (CAT) enzyme activities as well as capsaicin content (Islek *et al.*, 2016). Effects of several elicitors (yeast extract, salicylic acid and glycine) were reported to enhance the antioxidant (6-gingerol and 6-shogaol) and phenolic compound production from callus *Zingiber officinale* Rosc. (Ali *et al.*, 2018b). Anasori and Asghari (2009) reported light stimulated the secondary metabolites products such as gingerol and zingiberene in the callus culture of *Zingiber officinale*.

7.2 Genetic/Molecular Marker

India holds a significant position in globe with respect to the herbal drugs (Ganie *et al.*, 2015) and total estimated value of annual botanical trade in India is about US$10 billion (Singh *et al.*, 2003). Though the spices have high demand but natural habitat of spices has reduced day by day due to deforestation and industrialization and thereby supply of quality products to consumers have become a great challenge for the producers (Babu *et al.*, 2012). Hence inspection of the impurities as well as maintenance of the authenticity of plant samples (e.g. spices, traded medicinal plants) are considered as most desired parts to evaluate, promote and provide the quality, balance competition and consumer protection against excessive practices respectively (Dhanya and Sasikumar, 2010). But inadequate knowledge about the morphological features as well as lack of information regarding cultivar specific characters makes the discriminations of cultivars more complex (Kar *et al.*, 2014). Knowledge on genetic variation is prerequisite to study the evolutionary history of a species as well as to find out relationship among the species (Islam *et al.*, 2007; Das *et al.*, 2011). Besides study on genetic diversity within and among populations of a crop are essential to conserve the gene pool of that crop for future purpose (Mustafa *et al.*, 2005). On this context, genetic marker plays significant role to find out the variations among the different varieties of cultivars.

Marker

A marker is described as a mark to distinguish between two objects on the basis of some parameters. To study the diversity among different populations of a species, genetic marker should be polymorphic in nature. Polymorphism pointed out the alternative forms of a distinct phenotype in same population.

Genetic marker that requires to study the diversity/variation is classified as either-

i. *Morphological* marker studies the variations on the basis of morphological parameters; or
ii. *Biochemical marker* studies the variations at enzymatic level; or
iii. *Molecular marker* studies variation at genomic level. Among these markers, molecular marker is most useful because of their several advantages over morphological and biochemical marker. Each types of marker with respect to their advantages and disadvantages as well as role in spices are given below.

Morphological Marker in Spices

Though use of morphological marker is limited may be due to low polymorphism, low heritability, late expression and environmental sensitivity (Smith and Smith, 1992; Beyene *et al.*, 2005) but has been reported for several spices. It has been reported to study the diversity among 14 different populations of *Bunium persicum* (Chahota *et al.*, 2017); 20 accessions of Iranian *Cuminum cynimum* (Moghaddam and Pirbalouti, 2017); 25 different accessions of Indian turmeric (Bahadur *et al.*, 2016). Characterization of 56 accessions of *Capsicum chinense* (55 accession from Brazil and 1 from Peru) on the basis of 37 morphological variables was reported by Bianchi *et al.* (2020); while Cardoso *et al.* (2018) characterized 116 different accessions of *C. baccatum* from different regions of Brazil based on fruit traits (fruit color, shape, surface etc) as well as molecular marker (AFLP). Morphological parameters were also reported to characterize 15 genotypes of turmeric (Aarthi *et al.*, 2018). Use of morphological parameters has also been reported to study the diversity among 41 accessions of *Allium* from 13 species (Aryakia *et al.*, 2016). Luitel *et al.* (2018) studied variations among 47 accessions of *Capsicum chinense* Jacq. using 14 qualitative and 16 quantitative characters.

Biochemical Marker in Spices

Biochemical markers are the isozymes defined as enzymes having structural variations encoded by different gene with the same catalytic activity. The variations in isozyme are based on variations in amino acid content, and conformation of enzymes which are detected during electrophoresis followed by specific staining (Vallejos, 1983; Kumar *et al.*, 2009). Alloenzyme is the allelic variation of genes present in an enzyme. Hence the variations in alloenzyme is the products of different alleles instead of different genes of the enzyme that can be detected by monitoring the electrophoretic mobility during electrophoresis which arises due to point mutation in enzyme (Xu, 2010). Isozyme electrophoresis is chosen for its relative simplicity because it provides direct visualization of gene products (Brewer and Sing, 1970) and has ability to give a unique fingerprint for each genetically distinct clone (Lebot *et al.*, 1991; Criley *et al.*, 2006). These types of markers are codominant in nature (Govindaraj *et al.*, 2015), and do not need any information regarding the DNA sequence, probe. Inspite of these advantages, basic limitations are their limited availability as well as expression of protein/enzyme is specific to developmental stage of tissue (Winter and Kahi, 1995).

Use of isozymes has been reported to study the genetic diversity for several spices. Mondal *et al.* (2016) studied the diversity among 37 diverse chilli (*Capsicum annuum*) genotypes using the peroxidase polymorphisms while use of peroxidase (PO) and polyphenol oxidase (PPO) isozyme has been reported by Kumar *et al.* (2014) to find out the variations among 21 cultivars of chili; 4 different isozymes (acid phosphatase, estaerse, peroxidase and polyphenolic oxidase) used to study relationship among 20 varieties of ginger (Kar *et al.*, 2014).

Molecular Marker in Spices

Genetic marker is a gene or DNA segment having known location to chromosome which controls a particular trait (Nadeem *et al.*, 2018). Molecular/ DNA marker does not represent the gene of interest but it is a DNA segment at close proximity to the gene of interest and acted as “flags” or “marks”. Molecular markers are considered as effective tool in various fields of plant science (i.e. plant physiology, taxonomy, plant breeding and genetic engineering) because of their ability to detect the variations at nucleotide label of genome (Kesawat and Kumar, 2009; Zou *et al.*, 2011). They are more useful than other marker system (i.e. morphological/biochemical) to study the diversity among the plant varieties due to their direct access at genomic level (Kordrostami and Rahimi, 2015). Besides availability at all tissues as well as free from environmental influences make molecular marker most appropriate to study variation among the populations (Trindade, 2010). An ideal molecular marker should be highly reproducible, distributed throughout the genome, codominant, highly polymorphic and free from environmental effects (Joshi *et al.*, 1999). Allelic variations among the individuals of same species are results of either change in the number of tandem repeats at a particular locus (microsatellites) or insertion/deletion of a segment or single nucleotide polymorphisms (SNPs) (Mammadov *et al.*, 2012).

Classification of Molecular Marker

Molecular markers are primarily either monolocus that tags a single locus (e.g. RFLP, SNP etc.) or multilocus that tags with multiple locus (RAPD, ISSR, AFLP etc.) (Khlestkina and Salina, 2006). On the basis of mode of mechanism, they are broadly classified into three categories i.e. Hybridization based marker, PCR based marker and Sequence based marker.

Hybridization based marker was the first marker to be developed in 1980s and termed as Restriction Fragment Length Polymorphism (RFLP). Basic principle behind the polymorphism in DNA fragment relies on occurrence of number of restriction site within the sequence as well as variation in fragment length of restriction digested DNA. Visualization of banding pattern of restriction fragments in gel is based on principle of hybridization technique using the radiolabelled probe of known sequence (Bagali *et al.*, 2010). Though RFLP is highly polymorphic, reproducible, and codominant in nature but requires high amount of purified DNA as well as blotting technique make this marker system laborious and time consuming. Though RFLP has been reported to study the genetic diversity of other plant species but still it has not been reported in the spices. Hence till now most of the reports related to genetic diversity of spices is based on PCR based marker system i.e. RAPD, ISSR, AFLP, SSR etc.

Random Amplified Polymorphic DNA (RAPD) marker is a single random primer (10 base nucleotides) based marker system, has considered as the simple and cost-effective (Babu *et al.*, 2014). On the basis of primer length, RAPD is subdivided into either DNA Amplification Fingerprinting (DAF) for primer length less than 10 nucleotides or Arbitrarily Primed Polymerase Chain

Reaction (AP-PCR) for primer length about 15 nucleotides (Bardakci, 2001). In this technique, arbitrary primers are annealed to different complementary sites of genomic DNA in opposite direction and annealing within amplifiable distance permits amplification of target sequence using the Thermal Cycler. Similar to RAPD, Inter Simple Sequence Repeat (ISSR) is also PCR based marker system allows amplification of DNA segment present between two oppositely oriented identical microsatellite repeat regions using microsatellites primer (repeats of di-nucleotide, tri-nucleotide, tetranucleotide or penta-nucleotide) (Reddy *et al.*, 2002). Both of these markers have several advantages such as less expensive, easy to handle, less amount of DNA is needed to perform the reaction, no requirement of prior knowledge about genomic DNA sequence (Esselman *et al.*, 1999; Sheeja *et al.*, 2013). Because of these advantages, both of RAPD and ISSR have been reported to study genetic diversity for several spices e.g. between seeds of 19 anise landraces (*P. anisum* L.) (15 Turkish anise landraces of Turkey and four foreign landraces of Cyprus, Syria and France) (Giachino, 2020); 25 accessions of *Crocus* L. (Iridaceae) (5 accession of cultivated *C. sativus*, and 20 accessions belonging to wild species namely, *C. speciosus*, *C. cancellatus*, *C. haussknechtii*, *C. michelsonii*, and two unknown accessions from Iran) (Zarini *et al.*, 2019); 80 selected genotypes of *Zingiber officinale* Roscoe from North Eastern India (Baruah *et al.*, 2019); 24 diverse coriander genotypes from nine different geographical areas of the India (Choudhary *et al.*, 2019);4 different population of wild habitats of *Curcuma aromatica* from four different regions of Odisha (Parida *et al.*, 2018); 7 varieties of *Foeniculum vulgare* (FNL-41, 46, RF 101, 125,143, 178 and 205) (Grover and Malik, 2017);19 cultivars of turmeric (Basak *et al.*, 2017) etc. Though both of these markers are less laborious and easy to develop but major disadvantage is their dominant nature (Vijayan, 2005). In addition to that, RAPD marker has poor reproducibility and is unable to detect the reason behind the absence of band (https://www.ncbi.nlm.nih.gov/probe/docs/techrapd/). In compare to RAPD, ISSR is much more reproducible (Semagn *et al.*, 2006a) with more polymorphic (Reddy *et al.*, 2002; Hadian *et al.*, 2015).

Amplified Fragment Length Polymorphism (AFLP) is combined approach of RFLP and RAPD which permits selective amplification of restriction digested fragments of genomic DNA through the polymerase chain reaction (PCR) (Blears *et al.*, 1998). This mutilocus marker system follows multistep procedure involves- (a) restriction digestion of genomic DNA using two restriction enzymes and ligation of restriction fragments to site specific adapters, (b) pre selective PCR amplification of a subset of the restricted fragments using primer complementary to adapter sequence; (c) selective PCR amplification by adding selective bases to 3' end of primer to reduce further fragment number since amplifiable product depends on the complementarity between DNA fragments with nucleotides flanking the restriction sites and the selective nucleotides of the primer and finally (d) electrophoretic separation of amplified DNA fragments and detection of amplified fragments in gel (Paun and Schönswetter, 2012). AFLP has several advantages such as it requires less amount of DNA in compare to RFLP (Savelkoul *et al.*, 1999), more reliable (than ISSR and RAPD) (Costa *et al.*, 2016) and capacity to generate large numbers of marker fragments without having prior knowledge of genomic sequence (Maughan *et al.*, 1996).Inspite of such advantages, being a dominant marker AFLP is unable to distinguish the homozygote and heterozygote individual (Weeks *et al.*, 2000).

Till now several reports are available regarding application of AFLP to study the genetic diversity of spices. It has been reported to study genetic variability among a collection of 17 ginger clones (Varghese, 2018); 61 accessions of ginger from the ESALQ/USP's germplasm bank (representing 55 Brazilian ginger accessions collected from different geographical states and six Colombian ginger) (Blanco *et al.*, 2016). Molecular characterization using AFLP has been reported for 6 accessions from Iran, 2 accessions from Italian and along with 3 supplementary (one from Spanish, two from Italy) Saffrons (Torricelli *et al.*, 2019); 12 accessions of *Kaempferia* belonging to 9 species (Sihanat *et al.*, 2016).

Simple Sequence Repeats (SSR or STR), is short tandem repeated DNA are hypervariable in nature due to the changes in the number of repeat units (Tautz, 1989; Robinson *et al.*, 2004) has considered as most versatile marker. Polymorphism in SSR marker arises due to slipped-strand mispairing, or changes in length of repeats due to unequal crossing over and (or) subsequent replication, repair, or recombination errors (Levinson and Gutman, 1987; Nagy *et al.*, 2007). As a result, mutation in insertions or deletions of one or more repeat elements variations change in length of repeats (Wierdl *et al.*, 1997; Xu *et al.*, 2000). This variations in repeat number help SSR to study different aspects of genetics i.e. genotype identification, analysis of genetic diversity, phenotype mapping and marker assisted selection of crop plants. SSR is highly preferred by researchers for tagging / mapping traits of economic importance due to monolocus, multiallelic, highly polymorphic, co-dominant and cross species transferable nature (Jayaram *et al.*, 2016).

SSR marker has been reported to study the genetic diversity among 42 *Capsicum* genotypes of 11 species (Guzmán *et al.*, 2020); 90 accessions of *Perilla frutescens* from Korea and China (Park *et al.*, 2019); 5 natural populations of *Satureja bachtiarica* Bunge (Movahedi *et al.*, 2019); 53 Indian garlic accessions (Kumar *et al.*, 2019); 20 different chilli (*Capsicum* spp.) genotypes of Bangladesh (Sharmin *et al.*, 2018); 7 wild and 14 cultivated populations of *Aframomum corrorima* (Braun) P.C.M. Jansen (Chombe *et al.*, 2017). Besides development of SSR marker and followed by its validation in plant germplasm has been reported in several spices e.g. among 5 different genotypes of *Trachyspermum ammi* L. (Jogarana *et al.*, 2019); 30 black pepper accessions (Kumari *et al.*, 2019); 14 accessions of coriander cultivated in India (Tulsani *et al.*, 2020). Genomic simple sequence repeat (gSSR) markers developed from cumin variety GC-3 to study diversity analyses and cross-transferability in 30 cumin genotypes and 5 other species of family Apiaceae (Bharti *et al.*, 2018). Inspite of such advantages SSRs marker construction from genome is relatively expensive and based on genome-wide sequences. To counter these problems scientists derived sequence based molecular markers such as Expressed Sequence Tag (EST), Single Nucleotide Polymorphism (SNP).

Expressed Sequence Tag (EST)-SSRs are constructed from expressed sequence/coding regions of genome to reduce the manufacturing costs (Yang *et al.*, 2017). Though EST-SSRs are easy to construct by electronic search and highly transferable to related taxa but major disadvantage is the sequence redundancy that yields multiple sets of markers at the same locus (Chen *et al.*, 2015a). Development of EST-SSR markers followed by its validation has been reported for 35 genotypes of black pepper varieties and cultivars (Jose *et al.*, 2018); also reported to study the genetic diversity among 25 accessions of *Z. officinale* collected from different locations of India and check their cross transferability within the Zingiberaceae family (Awasthi *et al.*, 2017); reported to check the cross generic transferability from ginger (*Zingiber officinale*) to cardamom (Sakthipriya and Sabu, 2018); used to molecular identification of 8 elite cultivars and 88 accessions of in turmeric (Sahoo *et al.*, 2017) and diversity and relationship among 35 local cultivars of celery (Jethra *et al.*, 2017).

Single nucleotide polymorphism (SNP) is the most abundant marker and uniformly distributed throughout the genome. Polymorphism of SNP is based on mononucleotide variation within genome i.e. sequence variation due to change in only one nucleotide of genome (Qiu *et al.*, 2017). Since this variation is most common and stable, hence SNP is attractive marker system to study the diversity much more efficiently than other marker system. With respect to other molecular markers (described earlier), SNP is less labor intensive and less time-consuming, and the associated costs allow performing high-throughput genotyping. Though SNPs markers are biallelic, have lower information content than multiallelic SSR markers, but they occur at much higher density in genome, and have lower genotyping error rate (Šimić *et al.*, 2009). On the basis of these advantages, SNPs have been reported for several spices to perform different purposes e.g. it has been reported to study genetic diversity in a subset of 222 cultivated pepper (*Capsicum annum*) genotypes (Taranto *et al.*, 2016); 142 *Capsicum* genotypes from different geographical areas of Ethiopia (Solomon *et al.*, 2019); 190 *Capsicum* sp. accessions, including 183 of 5 cultivated species (*C. annuum*, *C. chinense*, *C. frutescens*, *C. baccatum*, and *C. pubescens*) and 7 of the wild form *C. annuum* var. *glabriusculum* (Pereira-Dias *et al.*, 2019).

At modern era of molecular biology, scientists are enormously trying to modify the marker system to improve the breeding programme as much as possible. To overcome the problems of conventional molecular markers, scientists developed several modified DNA sequence based markers. Among them Sequence Characterized Amplified Regions (SCAR); Start Codon-Targeted (SCoT); Cleaved Amplified Polymorphic Sequences (CAPS) and Sequence Related Amplified Polymorphism (SRAP) are significant and have gained importance to plant breeders due to their several uniqueness over conventional molecular marker system.

Sequence characterized amplified regions (SCAR): To counter the problems of RAPD, scientists developed another PCR based marker i.e. Sequence Characterized Amplified Region (SCAR). This marker system is highly reproducible, easy to use and locus specific (Kumar *et al.*, 2009). For SCAR, the primers are constructed from nucleotide sequence established in cloned RAPD fragment linked to a trait of interest (Bhagyawant, 2016). SCAR has high level of polymorphism may be due to higher annealing temperatures and longer primer sequence specificity (Kumla *et al.*, 2012; Yang *et al.*, 2013). SCAR markers are advantages over RAPD marker because of less sensitive to PCR reaction conditions, able to detect single locus and codominant in nature (Weng *et al.*, 1998). Due to these advantages, analysis through SCAR is more informative and straightforward and easy to perform. Use of SCAR has been reported to detect adulteration in 24 different food products of saffron (*Crocus sativus* L.) (Torelli *et al.*, 2014); authenticity and adulterants of *Lycium barbarum* (wolfberry) (Sze *et al.*, 2008). Development of SCAR marker in spices also reported to select disease resistance from the population for e.g. development of SCAR reported linked to powdery mildew resistance diease in chilli pepper (*Capsicum annuum* L.) (Wankhade and Wadikar, 2018), linked to *Phytophthora* root rot resistant plants in pepper breeding programs (Truong *et al.*, 2013); linked to genic male sterility of chili (Lee *et al.*, 2010).

***Start Codon-Targeted (SCoT)*:** As research proceeds, scientist tried to develop the gene-targeted markers instead of unidentified DNA markers to get more appropriate result. On this background, they developed start codon-targeted (SCoT) marker which is a PCR based marker system uses single primer targeting the short consensus conserved region flanking the ATG translation initiation codon (Xiong *et al.*, 2011). Though SCoT marker uses single long primer (18 nucleotides) than ISSR, SSR and RAPD but more polymorphic in nature (Feng *et al.*, 2018). In addition, it is simple, inexpensive, and highly polymorphic, shows extensive genetic information and primers are universal in plants (Guo *et al.*, 2012). Inspite of being dominant marker, these advantages make SCoT an attractive choice for plant breeders in current situation. Till now several attempts have been taken to use SCoT marker in spices. It has been reported to study the genetic polymorphism, cross transferability (CT) and genetic diversity among 54 different accessions of *Capsicum* species (Gupta *et al.*, 2019); to study the diversity among 51 accessions of Indian opium poppy germplasm (Srivastava *et al.*, 2020); 15 accessions of *C. annuum* from Cross River and Ebonyi States of Nigeria (Igwe *et al.*, 2019); 72 accessions of the Iranian poppy (*Papaver bracteatum* Lindl.) (Qaderi *et al.*, 2019); 6 fennel cultivars (*Foeniculum vulgare* Mill.) (Ramadan *et al.*, 2019); 39 Iranian black cumin landraces (*Nigella sativa* L.) (Mirzaei and Mirzaghaderi, 2017).

Cleaved Amplified Polymorphic Sequence (CAPS): This marker system is a combination of PCR and RFLP based on amplification of the DNA by PCR followed by digestion of the PCR product with restriction enzymes and finally polymorphism in amplified products are detected through electrophoretic separation followed by staining. CAPS has several advantages such as it is highly reproducible, specific, less laborious and time-consuming than RFLP (Elangbam and Misra, 2016). In addition, they are much more valuable than other markers (which are based on non-functional sequences) since they based on expressed genes (ESTs) (Semagn *et al.*, 2006a) along with co-dominant manner (Matsumoto and Tsumura, 2004). But till date no report available regarding application of CAPS in spices.

Sequence Related Amplified Polymorphism (SRAP): A PCR based marker which is used to amplify the coding regions of DNA with primers targeting the open reading frame (Robarts and Wolfe, 2014). This markers are highly potential over other marker system to study the molecular diversity, gene mapping and fingerprinting of genotype due to their multiloci and multi-allelic nature (Zaefizadeh and Goliev, 2009). This marker has several advantages such as it is simple, reliable, codominant in nature and able to target only coding region of genome (Li and Quiros, 2001). Use of SRAP were reported to study the genetic diversity of 24 different sumac (*Rhus coriaria L.*) genotypes (Sütyemez *et al.*, 2019); 16 genotypes of cumin (Bhatt *et al.*, 2017); 5 different *Salvia* species (Aghaei *et al.*, 2017); 28 saffron accessions collected from different regions of Iran (Babaei *et al.*, 2014).

7.3 Molecular Assisted Selection (MAS)

The principle objective of plant breeding is to improvement of crop in terms of several parameters such as yield, or qualities, and other traits having commercial significance to society (Moose and Mumm, 2008). Conventional plant breeding method is the common practice of plant breeders to create new varieties involves hybridization between two sexually compatible plant species followed by the phenotypic selection (Prohens, 2011). This method permits selection of superior genotype from the population on the basis of phenotypic expression of trait. Beside transfer of desirable gene/trait, this breeding method also allows segregation of the undesirable gene/trait(s) in next generations which are difficult to eliminate even through backcrossing method. In spite of having such limitation, conventional method plays significant role in conservation of wild germplasm, allows sexual hybridization between contrasting parental lines, and creates unique genetic variants (Werner *et al.*, 2005; Das *et al.*, 2017). Though traditional plant breeding method is considered as effective tools for crop improvements but it has several striking limitations-

i. Conventional breeding programme allows the breeders to incorporate the desirable trait into hybrid from parents of related species. Hence the modification of target trait is never go beyond the preexisting genetic potential of these species (Acquaah, 2015).

ii. Development of new variety through conventional approach is long time process (in most the cases it takes more than 8 years) and influenced by environment (Ibitoye and Akin-Idowu, 2011).

iii. Conventional plant breeding has become popular for breeding purpose may be due to its less complex analytical approach. In this method interpretation of result (selection of best from the population) is based on the phenotypic performance of trait in plant. But phenotypic performance is cumulative effect of genotype as well as environment. Hence phenotypic based selection results an imperfect prediction about the appropriate genotype. It makes the event (selection) more complex when the trait is polygenic in nature (controlled by multiple genes). So most of the phenotypic based selections are represent the collective effect of all genes and the environment (Dekkers and Hospital, 2002). So conventional method does not consider as effective tool to select the "best" from progeny population.

All these above mentioned problems related to the traditional breeding could be overcome by changing concept of "phenotypic based selection" to "molecular based selection". Molecular based selection through molecular marker is used to fulfill several purposes of breeding programme i.e. to recognize the cultivars, tested the authenticity of purity and study genetic diversity among the available germplasms, help in parental selection, study of heterosis and identification of genomic regions under selection (Collard and Mackill, 2008).

Molecular Breeding or Marker Assisted Breeding

Molecular breeding is "molecular marker based modern breeding technique" which is used as substitute of the phenotype selection to promote release of new improved germplasms (Cairns *et al.*, 2012). In this programme, use of molecular markers along with linkage map and genomics is common approach to develop and modify the trait of interest. This is termed as marker assisted breeding (MAB) (Jiang, 2013). MAB is also used in several breeding techniques including Marker-Assisted Selection (MAS), Marker-Assisted Backcrossing (MABC), Marker-Assisted Recurrent Selection (MARS), and Genome-Wide Selection (GWS) or Genomic Selection (GS) (Ribaut *et al.*, 2010). These are briefly described here below:

Marker Assisted Selection (MAS): MAS involves the utilization of molecular marker which is adjacent to target loci to be screened. This selection method acts as alternative option or to assist phenotypic screening (Wijerathna, 2015).

Marker Assisted Backcrossing (MABC): With the help of molecular markers, in this method it is aimed to introgress a locus in a superior background (recurrent parent) lacking the trait retaining the essential characteristics of the recurrent parent (Collard and Mackill, 2007).

Marker Assisted Recurrent Selection (MARS): In order to develop the best performing genotype for a complex trait controlled by multiple genomic regions, MARS is followed to select those genomic loci and transferred into the new background within a single or across related populations (Ribaut *et al.*, 2010).

Genomic wide selection (GWS): GWS estimates effect of the all genome wide marker at a time based on a reference population known phenotype (Bhering and Xu, 2019).

Marker Assisted Selection (Mas) in Plant Breeding

Applications of molecular markers in cultivar development have considered as emerging field of plant breeding/biotechnology involves "Marker Assisted Selection (MAS)". Basic purpose of MAS in breeding programme is to recognize the markers which are tightly attached / linked to a trait controlling gene and use this association (marker linked trait/gene) for development of new varieties and breeding lines through traditional breeding techniques (crossing, backcrossing, self pollination, and selection) (Leonova, 2013). In addition to this, selection procedure can be performed at early stage of development which is free from environmental impact hence MAS is reliable, time and resource saving method in compare to the conventional method (Torres *et al.*, 2010). Molecular marker in MAS helps to fulfills three basic purpose of breeding – i.e. uses to trace favorable allele(s) (dominant or recessive) across generations in order to accumulate favorable alleles; helps in selection of the most appropriate individuals among segregating progenies (based on the allelic composition of a part or of the entire genome) and also breaks the possible linkage of favorable alleles with undesirable loci (Francia *et al.*, 2005). So applicability of MAS is most suitable for those types of traits which are either complicated to detect or costly to phenotype or controlled by multigenes as well as it helps to identify and maintenance of recessive alleles of genes in a population (Leonova, 2013). Though use of MAS is advantageous over phenotypic selection for large population as well as applicable for traits having relatively low heritability but extra genetic gain provided by MAS, with respect to phenotypic selection has reduced when several successive cycles of selection were considered, and hence MAS could become less efficient than phenotypic selection in the long term (Hospital *et al.*, 1997).

Important requirements for MAS: Important requirements for effective MAS involves use of reliable marker system, availability of quick DNA extraction method, high throughput marker detection system, knowledge on marker-trait association and construction of genetic map, interpretation of result (Jiang, 2015). Significance of these requirements in MAS have discussed below briefly-

- *The marker system using for MAS should be reliable in nature*: The relationship between molecular marker and gene of interest/trait is available in any one from the following forms- either– marker present (i) within the gene of interest or (ii) in linkage disequilibrium with the gene of interest throughout the population (for this condition close contact/association is mandatory between marker and trait of interest) or (iii) in linkage equilibrium with the gene of interest throughout the population. Among these three conditions, condition (i) is most suitable and efficient for the MAS but very rarely available in nature (Babu *et al.*, 2004). Marker close to trait of interest is most suitable for MAS since it reduces the scope of the recombination between targets with the marker. Hence molecular marker which is highly polymorphic for the genotypes, tightly linked with the trait of interest (distance should not be more than 2 cM), cost effective as well as PCR based nature has considered as most suitable for rapid screening of large populations (Mohler and Singrün, 2004; Nogoy*et al.*, 2016).

- *Access to quick DNA extraction method and high throughput marker detection system*: In breeding programme, breeders deal with plant population to screen the trait of interest. Screening of population through molecular marker approach involves DNA extraction at early stage which becomes time-consuming and laborious for screening large population. Hence availability of quick DNA extraction method as well as high throughput marker based detection method are most desirable to collect the DNA sample and evaluate the large population within short time respectively. PCR based markers need low amount of DNA as well as easy to construct so these markers can be efficient for screening the large population. Markers which are codominant, tightly linked, and highly polymorphic in nature usually good for MAS (Wijerathna, 2015). Besides, screening techniques having high reproducibility across laboratories, user friendly and cost effective nature put extra advantages to selection procedure and makes the MAS most effective. Microsatellite markers, also termed sequence-tagged microsatellite site (STMS) or simple sequence repeat (SSR) markers, is presently considered as the most complete tool for MAS (Mohler and Singrün, 2004). SSR is most suitable may be due to its unique nature (i.e. co-dominant, highly polymorphic, available throughout the genome and highly reproducible) ((Described in details in 7.2).
- *Breeder should have knowledge to construct the linkage map and marker-trait association*: Linkage map provides the information about the relative distance between the genes. Requirement of mapping is essential for MAS due to their ability to recognize the molecular markers or candidate genes associated with traits. The breeders must have enough information about the markers and trait association as tightly linked markers are useful in mapping.
- *Breeder should have the ability to collect the date efficiently and interpretation ability.*

Basic steps of MAS: On the basis of such above mentioned requirements, breeders conduct the basic steps of MAS. Most common way to conduct MAS involves selection of parents on the basis of contrasting character followed by hybridization and selection of desirable progeny using linked marker with respect to trait of interest. Basic outline of MAS procedure is given below (Fig. 7.3.1).

1. Selection of suitable parents	• Selection of contrasting trait (e.g. Disease resistant and Diseases susceptible) • Identification of marker linked to the target trait (e.g. linked to disease resistant)
2. Mapping population	• Generation of F1 population. • Selfing to produce F2 generation.
3. Analysis based on molecular approach	• Extraction of DNA from sample of F2 population • Use of marker and amplify the product. • Qualitative analysis through Agarose gel electrophoresis • Selection of plant having desirable trait (e.g. Disease susceptible against disease resistant on the basis of absence of band).

Fig. 7.3.1: Basic Outline of Marker Assisted Selection (MAS)

Application of marker assisted selection (MAS) in plant breeding covers several aspects of plant breeding. These aspects can be divided into five different categories (Collard and Mackill, 2008)-

A. Marker-assisted evaluation of breeding materials

B. Marker-assisted backcrossing

C. Gene pyramiding

E. Early generation selection

F. Combined MAS

All of above mentioned aspects are discussed below with special reference to spices-

A. **Marker Assisted Evaluation of Breeding Materials:** In traditional breeding programme, choice of parental germplasms and their assessment are carried out through visual selection followed by morphological analysis. Assessment of plant samples (e.g. spices, traded medicinal plants) are most needed part to evaluate, promote and provide the quality, balance competition and consumer protection against excessive practices respectively (Dhanya and Sasikumar, 2010). But inadequate knowledge about the morphological features as well as lack of information regarding cultivar specific characters makes the discriminations of cultivars more complex (Kar *et al.*, 2014). Knowledge on genetic variation is prerequisite to study the evolutionary history of a species as well as to find out the relationship among the species (Islam *et al.*, 2007; Das *et al.*, 2011). In addition, phenotypic expression is influenced by environment. Hence judgment of the efficient genotype through visual observation undoubtedly makes the breeding procedure unrealistic. On this context genetic marker plays significant role to find out the variations among the different varieties of cultivars at molecular level (Kar *et al.*, 2014). DNA markers in MAS play an important role to monitor the favorable/desirable allele(s) (dominant or recessive) across generations and identify the most suitable individual(s) from the segregating progeny on the basis of allelic composition across a part or the entire genome (Obayemi and Micheal, 2016). Use of marker to find out the genetic diversity as well as genetic similarity spices have already been reported with respect to different types of PCR based or sequence based molecular markers. Please see 7.2 for details.

B. **Marker-assisted backcrossing (MABC):** One of the major goals of plant breeding is to transfer of one or more genes from a donor into the recurrent parent and recovery of the recurrent parent genome. This objective is traditionally carried out by backcrossing method. Rate of recovery of recurrent parents depends on numbers of backcrossing breeding [e.g. the recovery rate of recurrent parent will be 99.2% after 6th generation of backcrossing (after t generation of backcrossing, recovery rate of recurrent genome is 1 – (1/2)^t+1] (Babu *et al.*, 2004). So large numbers of backcrosses are essential to allow introgression of desire trait which makes recovery of recurrent genome time consuming. Besides successive backcrosses also allow transfer of other genes (undesirable) adjacent to the gene of interest from the donor parent (linkage drag) to progeny (Giaever *et al.*, 2017). All of these problems could be eliminated through marker-assisted backcrossing (MABC) method. MABC is most convenient process to introgress a major gene or quantitative trait locus (QTL) into a popular variety. Recovery of recurrent parent through this method is achieved within short time with respect to conventional backcrossing method. In addition, this method has several other advantages i.e. it minimizes the transfer of undesirable one (linkage drag); free from environmental influences and provides more accurate selection (Hasan *et al.*, 2015). In this method three different levels of selection are done to perform three different objectives. At initial level of selection, molecular marker allows simple and efficient selection of the target locus for which direct selection is difficult ('foreground selection') (Frisch *et al.*, 1999) followed by 'recombinant selection' involves minimization of donor chromosomal part having the target locus and finally 'background selection' allows recovery of the recurrent parent by selecting backcross lines with a higher proportion of recurrent parent genome (Vu *et al*., 2012). Successful application of MABC is depending upon several factors such as population size of each backcross generation, closeness between the markers and the target gene, and number of background markers to be used (Hasan *et al*., 2015). Molecular marker which is located directly within the gene of interest (direct markers) is most suitable for MABC, is termed as gene assisted selection (Dekkers, 2003; Semagn *et al*., 2006b). Reports on application of marker assisted backcrossing in spices are very less. Through this method introgression of the heat shock protein (Hsp70 and sHsp) genes from heat-tolerant AVPP0702 (*C. annuum* L.) into the genetic profile of heat sensitive Malaysian elite chilli variety Kulai (*Capsicum annuum* L.) has been reported by Usman *et al*., 2018. Similarly, introgression of heat-tolerant gene (Hsps) into Kulai from donor AVPP0702 has been reported by Usman *et al*. (2020). They found 97% recovery of recurrent parent genome after BC3F2 and finally developed 12 improved heat tolerance chili line. Major problem to construct the CMS line for sweet pepper is unavailability of a restorer-of-fertility (*Rf*) allele. Lin *et al*. (2015) successfully reported introgression of *Rf* locus of hot pepper (AVPP9905) to sweetpepper BC_4F_2 genotype through marker assisted backcrossing method using a SCAR marker. This method has been reported to transfer the favorable alleles at four (4) major quantitative trait loci controlling resistance against *Phytophthora capsici* (responsible for root rot and shoot blight in pepper) from a small fruited pepper into a bell pepper recipient line molecular (Thabuis *et al*., 2004).

C. **Gene Pyramiding:** Plants in field are susceptible to several biotic and abiotic stresses which interfere with the yield potential of plant. In addition, attack of pathogen creates the situation more complex for the susceptible varieties. Uses of agrochemicals are common practice to reduce the infection but create health hazard as well as damage the soil health. In this context another scheme of MAS i.e. Gene pyramiding is an efficient method to induce durable disease resistance in crops. Multiple gene/or quantitative target loci from different parents are incorporated into a common background (Malav *et al*., 2016). Gene pyramiding is complicated or nearly impossible through conventional method due to intra- and intergenic effects of genes controlling disease resistance and linkage with undesirable traits that is difficult to break even with many generations of backcrosses (Young and Tanksley, 1989; Dokku *et al*., 2013). Gene pyramiding scheme involves two phases - at initial phase basic objective is to collect all the target genes into a single genotype to make the root genotype (pedigree step) while in second part involves fixation step which permits the fixation of target gene in its homozygous condition. In gene pyramiding though pedigree step may be same but fixation phase can be conducted using several methods (Servin *et al.*, 2004). Application of this scheme of MAS has been reported for very limited spices. A pyramiding scheme against Potato virus Y (PVY), Tomato spotted wilt virus (TSWV) and Pepper mild mottle virus (PMMoV) have been developed in sweet Charleston pepper line 'Y-CAR' using molecular markers and biological assays (Özkaynak *et al.*, 2014).

D. **Early generation selection:** In compare to conventional breeding, one of the major advantages of MAS is ability to screen the difficult traits at the seedling stage (indirect selection), therefore it is speeding up the selection process and helps in development of traits which are difficult to develop using the traditional plant breeding method. (Varshney *et al.*, 2009). This protocol reduces the experimental cost as well as time of the experiment (Slater *et al.*, 2013). CAPS marker tightly linked to tomato spotted wilt virus resistance gene has been developed for pepper (Moury *et al.*, 2000). Similarly, a SCAR marker linked to powdery mildew disease resistance for pepper has been developed by Wankhade and Wadikar (2018). STS marker to track turmeric (*Curcuma longa* L.) resistance against rhizome rot caused by *Pythium aphanidermatum* has been developed by Kar *et al.* (2014a). All these markers are safe and efficient to identify susceptible germplasms at an early stage to decrease the effect of disease on the overall productivity of the crop.

E. **Combined MAS:** In this method both of phenotypic selection as well as marker based selection is coexist. This is applicable when the objective is to achieve the maximum genetic gain, involves use of imperfect marker and reduces population sizes for traits where marker genotyping is cheaper or easier than phenotypic screening.

7.4 Transgenic Breeding

One of the age old practices of human civilization is "agriculture". Though the principle objective of agriculture is related to food security but during recent past agriculture has been badly impacted due to change in global climatic conditions. Besides, human activities such as uncontrollable deforestation and industrialization have reduced the area of cultivated land. As a result, the gap between demand and supply of food has increased. Like other crops, spices are also facing the same problem. Modern interventions have been introduced for "high yield oriented agriculture". The synthetic chemical intensive "high yield oriented agriculture" creates several problems on human as well environmental health. Disease susceptibility, unavailability of resistant varieties, and post harvest losses are major constraints for optimum productivity of spices (Babu *et al.*, 2013). Hence cultivation of high yielding or stress/disease resistance variety is the most viable option to achieve this goal. For this reason researchers have shifted their research interest from crop production to crop improvement.

Hybridization approach is common tool of the plant breeders to improve the plant characteristics. Since this method helps the breeders to introduce the target trait into hybrid from parents of related species, so the alteration of target trait never goes beyond the preexisting genetic potential of these species (Acquaah, 2015). Besides this method helps in the transfer of unwanted gene/ chromosomal segments as well as may leads to sterility due to adverse genetic effect (Khan and Liu, 2009). Development of the new variety using traditional methods is time taking process and affected by environment as well as depends on nature of the parental germplasms. Biotechnological approach "transgenic breeding" has considered as emerging technique to achieve the goal of "crop improvement". Most significant advantage of transgenic breeding over conventional breeding involves its wide applicability i.e. permits transfer of gene between two distantly related species (Bhojwani and Razdan, 1996). Advancement of genetic engineering in transgenic breeding allows precise transfer of gene across the gene pool (Visarada *et al.*, 2009).

Hence this technology omits the "intergenic" barrier as well as permits precise transfer of gene of interest into the target. Taking these advantages, scientists have focused on improvement for spices and tried to develop "transgenic spices" using this technology but reports regarding this are very less. Before going to details on spice improvement through transgenic breeding, basic principle and methodology of transgenic breeding are discussed below-

Genetic Engineering: Basic of Transgenic Breeding

Genetic engineering in common view facilitates manipulation of nucleic acid. Alteration in genetic makeup of an individual through biotechnological intervention is the basic application of genetic engineering. Principle stages of genetic engineering involves isolation of the gene of interest from source and ligation of the gene of interest with the carrier DNA (Vector) to form the recombinant DNA construct and finally transfer it to host. Biotechnologists are trying to modify the genetic background of host plant so that it will withstand the diseases or (biotic/abiotic) stresses.

Transgene and transgenic breeding

Through the genetic engineering process when gene introduces to the host it may be either *cisgene or transgene* which are supplied from same or unrelated species respectively (Rani and Usha, 2013). The nature of alterations involves either insertion/deletion of DNA sequence or replacement with exogenous sequence (Lanigan *et al.*, 2020). The basic objective of transgenic integression is to create unique characters in plant which does not exist earlier. Transfer of the gene may be carried out either through vertical or horizontal mode. Vertical gene transfer involves transfer of gene from parent to offspring while horizontal gene transfer allows the transfer of gene between two organisms (related or unrelated species) (Ponti, 2005). The plants having *transgen*e is termed as transgenic/genetically modified plant. The basic procedure of transgenic breeding has described below-

i. **Isolation of gene of interest for a particular character:** The most crucial step of the transgenic plant production starts with selection of the gene that is responsible for a unique trait and identification of its location from the chromosome of donor plant (Abbas, 2018). Various studies are prerequisite prior to selection followed by isolation of that gene for introduction/improvement of particular trait into/of recipient. Study on the additive effect as well as impact of that gene/trait to other genes/traits is an essential part prior to selection of the gene/trait of interest and plays a significant role to identify the target.

ii. **Construction of *trasngene*:** After isolation, the gene of interest is subjected to several modifications to form the suitable gene construct prior to transformation stage. In this phase the coding sequence of gene are modified by using a promoter, enhancer, and terminator. Use of promoter is essential for the expression of transgene while the levels of expression can be monitored by using this enhancer and terminator sequences (Visarada *et al.*, 2009).

iii. **Transfer of gene construct/transgene to host:** In this stage, the modified gene construct/transgene transfers to the plant using suitable DNA delivery method. This method is termed as transformation. Transformation in plant may be carried out either through(i) indirect / vector-dependent method involves transfer of gene through vector or (ii) Direct DNA transfer method involves direct uptake of naked DNA by the plant cells via physical method (involves gene gun electroporation, microinjection etc.) or chemical based method respectively (Gupta *et al.*, 2013). Among these gene transfer methods, most widely used indirect method of gene transformation for plant is carried out through *Agrobacterium tumefaciens*, the natural genetic engineer and directly using the 'gene gun' (Southgate *et al.*, 1995; Key *et al.*, 2008). But both of these methods have some striking limitations. Gene transfer in monocot is difficult through *Agrobacterium tumefaciens* because this bacteria has natural ability to transfer the gene only to the dicot plants and presence of wound in host plant is prerequisite for bacterial infection. Though gene gun method overcomes these limitations but ineffective for transfer of large fragment (since transfer of more than 10 kb DNA is difficult because large fragments can be destroyed during the bombardment or adhere poorly to the metal particle resulted improper/unorganized integression to cell) (Shou *et al.*, 2004) as well as integression of multiple identical gene within the genome promotes gene silencing (Anami *et al.*, 2013). In addition to the transformation method, there are several factors need to be considered for effective transformation.

iv. **Selection of efficient transforming cells and regeneration of transformants:**After efficient transformation either through physical method (biolistic gun-mediated) or indirect method (*Agrobacterium* mediated method), the recipient tissues are transferred into growth supporting media for a short time prior to transfer in a similar medium containing selection agent (herbicide or antibiotics) (Moeller and Wang, 2008). Since the frequency of successful transformation usually is very low, so selection of transformed cells is very crucial prior to regeneration of whole plant. Most of the cases the antibiotic resistance gene has been reported in gene construct to indentify the transformed cells against the non-transformed cells. After that the transformed cells are selectively cultured to regenerate whole plants over time under controlled condition. The presence of transgene in the target organism of T0 generation may be confirmed by different techniques like PCR, Southern blotting etc whereas the target trait expression in the host may be checked by western blotting, ELISA and RT-PCR techniques (Passricha *et al.*, 2016). Prior to final approval from authority, multistage evaluation of transgenic organisms are must needed to clarify several aspects of transgenic breeding such as - (a) to know the number of inserted copies into host (presence of single intact copy of transgene without having negative effect on host is only desirable); (b) desirable phenotypic expression (depending upon tissue specificity and growth stage of the plant life cycle); (c) steady and reliable expression throughout several generations regardless of other conditions (environment and stresses, genetic background etc.) (Mumm, 2013).

Transgenic Breeding in Spices

The application of transgenic breeding in spices is very less with respect to other common crops. With respect to other transfer method, *Agrobacterium* mediated gene transfer method has widely been reported for several spices to conduct efficient mediated transformation using kanamycin as selection agent (Niu *et al.*, 1998; Suma *et al.*, 2008; Varghese and Bhat, 2011; Mahato *et al.*,

2018). In spices, most of the reports related to regeneration of transformed plantlets are carried out through cocultivation of leaf discs of plant with the *Agrobacterium tumefaciens* in suitable media. The most of the research has been conducted in chili may be due to its significant position in spices (king of the spices).

Stress Resistance: Abiotic stresses are responsible for reduction of 70% crop yield globally. Stress is one of the major constraints for optimum productivity of spices. In spices, several attempts have been taken to develop transgenic plant against the stresses (salinity, drought etc). DNA helicase plays significant role in various molecular mechanisms. Besides it involves in myriad cellular processes that impart/promotes intrinsic tolerance against abiotic stresses in plants. Keeping view on this significant role against stress, role of Pea DNA helicase45 (PDH45) has been studied by Shivakumara *et al.*, 2017. They found PDH45 was effective to provide resistance against salt, and moisture in chili (*Capsicum annuum* L.). Accumulation of reactive oxygen species (ROS) due to salt stress is common phenomena in plant. Presence of *osmotin* gene has negative impact on this ROS and over expression of *osmotin* protein protects the cell from the ROS by preventing its accumulation within cell. Based on this principle, salt resistance chili using overexpression *osmotin* gene has been reported by Subramanyam *et al.* (2011). They developed transgenic salt tolerance chili based on the ectopic expression/over expression of the *Nicotiana tabaccumosmotin* gene and found high yield at high salt concentration with respect to control (normal/ non transgenic plant) i.e 3.32 kg chilli pepper fruits/ transgenic plant in compare to no production of non transgenic at 300 mM NaCl concentration. The *SbNHX1* gene cloned from *Salicornia brachiate,* an extreme halophyte, has been transferred in cumin (Pandey *et al.*, 2016). This gene involves in the compartmentalization of excess Na^+ ions into the vacuole and maintenance of ion homeostasis. They found overexpression of *SbNHX1* gene was also responsible for lower electrolytic leakage as compared to wild type plants under salinity stress. Similarly, from wheat, TaNHX2, an Na^+/ H+ antiporter gene, has been introduced to induce salinity tolerance in chilli pepper (*Capsicum annuum* L.) by Bulle *et al.* (2016). They reported survivality of transgnic chili at salt stress condition (200 mM NaCl) and found higher levels of several biochemical components (proline, chlorophyll, superoxide dismutase, ascorbate peroxidase) with respect to control plants (wild type) under salt stress conditions. In addition to stress, insect resistant chili has been developed by Zhu *et al.* (2015). They reported a gene pyramiding scheme to developed drought and insect resistance in chilli and found significance improve in survivality of transgenic chili under drought stress i.e. (61.1%) by introducing *Arabidopsis thaliana* homodomain-leucine zipper transcription factor. In addition, presence of *Cry2Aa2* gene from *Bacillus thuringiensis* enhanced the resistant against *Prodenia litura* which promoted 81.3% mortality after 5 days of infection.

Disease resistance: The use of virus coat protein is common approach to control plant viruses. Using this approach, disease resistant chili has been reported for chili pepper (*Capsicum annuum* var. Longunt) by Cai *et al.* (2003). They developed virus-resistant chilli pepper by transferring genes of the cucumber mosaic virus (CMV) and tobacco mosaic virus (TMV) coat protein (CP) through *A. tumefaciens* to chilli pepper (cultivar 8212). The transfer of gene for chilli leaf curl virus coat protein has been reported to generate chilli leaf curl virus resistance transgenic hot pepper by Raghunathachari *et al* (2014). Similarly expression of antifungal protein by gene provides resistance to plant against the fungal diseases. One of such gene reported in black pepper spices involved PR 5 (pathogenesis related) gene '*osmotin*' to delay infection and decreased spread of foot rot disease (causal agent *Phytophthora capsici*) (Sinoj *et al.*, 2014). Introduction of *Autographa californica baculovirus chitinase* gene through *Agrobacterium*-mediated transformation in chilli resulted tolerance against Anthracnose disease (causal agent *Colletotrichum* sp.) (Mythili *et al.*, 2015). A different approach i.e. host induced gene silencing (HIGS) approach has been developed by Mahato et al., 2020 to target the *Colletotrichum gloeosporioides COM1* (*CgCOM1*) using *CgCOM1*-RNAi construct. Through *Agrobacterium*-mediated transformation they found nearly complete degradation of *CgCOM1* transcripts via HIGS in the transgenic lines of chili and tomato. Transgenic turmeric using particle bombardment method followed by selection of transgenic plants reported against the herbicide was reported by Shirgurkar *et al.* (2006). They found the plant having transgene with *basta* had high resistance against the glufosinate.

In addition, several genes from spices have been reported as effective to gain resistance against several diseases for non spices crops. For example, candidate root-knot nematode resistance gene (designated as CaMi) from the resistant pepper line PR 205 was found effective to induce nematode resistance in tomato (Chen *et al.*, 2007);Over expression of *CaMBF1* (isolated from *Capsicum annuum*) reduced tolerance to cold and high salt stress during seed germination and post-germination stages of *Arabidopsis* (Guo *et al.*, 2014); in contrast over expression of ZoCDPK1 (a Calcium-dependent protein kinases) gene from ginger was reported to induce tolerance to salinity and drought stress in *Nicotiana tabacum* (Vivek *et al.*, 2013).

7.5 Omics

Crop improvement, the one of the major fundamental objectives of plant breeding is based on crop productivity and quality improvement. Availability of recently developed modern techniques of molecular biology help the breeders to achieve these targets by introducing the novel genes as well as desirable traits into plant (Hesse and Höfgen, 2001). Though genome sequence of organisms help the researchers to identify the functional product (protein having known function) but problem arises for the unknown protein having unknown function. Hence research interest in current trends of molecular biology involves to find out the function of these unknown proteins at both of molecular (biochemical and biophysical) and cellular level (Zhang and Kim, 2003). To overcome these problems, scientists developed the concept of "*Omics* Technology" which in broad sense helps in detection of the genes, mRNA, proteins and metabolites from the biological sample with an unbiased manner (Horgon and Kenny, 2011).

Central Dogma of Molecular Biology

Francis Crick (1958) proposed the concept of central dogma of molecular biology to represent a clear cut relationship between the genes with its product. It provides an idea about unidirectional flow of biological information from DNA (gene) to its functional product (protein). DNA replicates within nucleus to produce daughter strands which later converts into functional product via an intermediate called mRNA. Conversion of DNA to mRNA is known as transcription while conversion of mRNA to protein (polypeptide chain) is termed as translation. Transcription and translation take place in nucleus and ribosome respectively.

Significance Omics in Molecular Biology

The word "*Ome*" derived from Greek literature to define "*body*" but in biological science it is an object which may be used to describe a character/feature/wholeness. The word "*Omics*" represents the study of the one "*Om*e". Hence *Omics* facilitates the interdisciplinary study among the different-*omes* with the help of modern biological tools i.e. bioinformatics and computational biology (Jazayeri and Torres, 2017). The core parts of *Omics* technology is composed of gen*omics*, transcript*omics*, prote*omics* and metabol*omics*, are commonly used to analyze the genome,transcriptome, proteome and metabolome of cells and tissues of an organism respectively (Schneider and Orchard, 2011). Though the *Omics* study is based on the understanding of the complex interaction between the genetic constituent, genes, proteins, and metabolites, but is result of application of analytical methods, such as bioinformatics, computational analysis, etc., and many other disciplines of biology (Ali *et al*., 2019). Due to these advantages of *Omics*, this multidisciplinary approach is an unavoidable part of modern research. We can categorized the application of omics into four parts for this purpose i.e. genomics, transcriptomics, proteomics and metabolomics in spices. All of these have described in next four parts briefly with special reference to species.

7.5.1 Genomics in Spices

Genome of an organism highlights the haploid chromosome content. Hence study of the genome is termed as "genomics". Genomics provides several information to researchers regarding gene content, genetic organization, gene arrangement (mapping) of organism as well as their role in metabolic process using gene mapping and DNA sequencing method (Shalini *et al*., 2018). Though sequencing of genome has revolutionized the modern biology by providing the crude information about the organism but unable to detect the the roles of molecules within cells. This problem could be solved by the "Structural Genomics" approach involves use of the high throughput techniques along with biological assays (Skolnick *et al*., 2000). While another approach of genomics i.e. "Functional Genomics" highlights the function and gene interactions (Bunnik *et al*., 2013).In contrast, the comparison between the genomes of different organisms leads to development of " Comparative Genomics" which helps to find out the evolutionary relationship between organisms as well as genetic differences and similarities within and between species (Sivashankari and Shanmughavel, 2007). Due to these significant contributions, genomic study has become emerging field since last decade for spices. Though genome sequencing is common approach for study the genome but till date reports reading this in spices are very less. In spices most of the cases it is restricted for chilli/pepper. Full genome sequence has been reported by Qin *et al*. (2014) for cultivated pepper variety Zunla-1 (*C. annuum* L.) as well as Chiltepin (*C. annuum* var. *glabriusculum*); while Kim *et al*. (2014) reported whole-genome sequencing and assemble the genome sequence of the hot pepper genome (Mexican landrace of *Capsicum annuum*cv. CM334). With respect to whole genome sequencing, report on sequencing of chloroplast genome is much more common in spices. Chloroplast genome sequencing has been reported for *Mentha canadensis* chloroplast by Huaizhu *et al*. (2020) presence of 134 unique genes (composed of 89 protein coding genes, 37 tRNA genes and 8 rRNA genes); while annotated chloroplast genome sequencing (125 genes with 86 coding sequences, 38 tRNA genes, and 8 unique rRNA sequences) has been reported for *Curcuma longa* by Windsor *et al*. (2019). Complete chloroplast genome sequence has also been reported for Bolivian wild chili pepper by Sebastin *et al*. (2019). They found that it was made up of 113 unique genes (among which 79 genes were protein coding genes, 30 genes for tRNA genes, and 4 rRNA genes) by Raveendar *et al.* (2017) for *C. chinense* from where they reported presence of 117 SSRs. In case of ginger, development of chloroplast genome sequence reported form Alpine ginger species i.e. *Roscoea humeana* (Zingiberaceae) (Zhu *et al.*, 2019) and *Roscoea tibetica* (Zou *et al.*, 2020).

Another approach of genomics study involves construction of gene mapping. On this aspect, several reports are available for spices but most of them are related to *Capsicum* sp. Mapping of *ms1* gene occupied 869.9 kb region of chromosome 5 has been reported for *Capsicum annuum* L. by Jeong *et al.* (2018), while mapping of anthrocyanine biosynthesis for the same plant species has been reported by Liu et al. (2020). They mapped *Ca3GT* gene having 110.5-kb region through a map-based cloning strategy for this purpose. Fine mapping of quantitative trait loci (QTLs) having resistant against anthracnose (a fungal disease) has been developed by Zhao *et al.* (2020) for *Capsicum chinense* 'PBC932'. They found the major QTL resistant to *Colletotrichum scovillei (should be in italics)* interval to a 164 Kb region made up of five genes. Besides a novel strong candidate gene *CA00g82510* for the male fertility restoration locus *CaRf032* in *Capsicum annuum* was identified by genome re-sequencing and recombination analysis (Zhang *et al.*, 2020). Cloning and characterization of a *CcMYB* gene (involves in capsaicine synthesis of *C. chinense*) has been reported by Zhao *et al.* (2018). In addition, characterization and molecular cloning of several genes are reported form turmeric e.g. *Phytocystatin* (Chan *et al.*, 2014); *CHS* gene family (Wannapinpong *et al.*, 2015); *MAP kinase* genes (Nanda *et al.*, 2014).

Molecular marker is commonly used to detect the variations or find out the similarity among the germplasms. For this reasons several attempts has been reported to develop molecular marker from the genome sequence of different spices. On this aspect development of SSR marker have been reported for several spices e.g. 21 polymorphic microsatellites were identified from turmeric (*Curcuma longa* L.) and reported to screen 30 accessions of turmeric (Senan *et al.*, 2013); while an genome wide identification of microsatellite approach has been reported by Kumari *et al.* (2019) for black pepper. Among the identified SSRs, 50 SSRs were used to check the variations among diversity of 30 black pepper accessions. Development of microsatellite has also been reported for *Elettaria cardamomum Maton* by Cyriac et al. (2016). In addition to microsatellites, development of single nucleotide polymorphisms (SNPs) has been reported for pepper by Taranto *et al.* (2016). Genome-wide identification of Insertion/Deletion polymorphisms (InDels) in *Capsicum sp.* was performed through comparing whole-genome re-sequencing data from two *Capsicum* accessions (*C. annuum*cv. G29 and *C. frutescen*s cv. PBC688) with the reference genome sequence of *C. annuum* cv. CM334 by Guo *et al.* (2019). Genome wide discovery single nucleotide polymorphism (SNP) marker related to powdery mildew (PM) resistance in two pepper varieties (*Capsicum baccatum* (PRH1) and *Capsicum annuum* (Saengryeg) has been developed by Ahn *et al.* (2016). In recent time use of molecular marker to detect the adulations in spices has replaced with DNA barcode. Beside the molecular marker, use of DNA barcoding has gained popular due to its ability to identify the taxonomic origin as well as authencity of germplasms. On this aspect, use of barcoding has been reported for several spices (Parvathy *et al.*, 2015; Jiang *et al.*, 2014; Aghighiravan *et al.*, 2019).

7.5.2 Transcriptomics in Spices

Transcriptome covers the total RNA content of cell at a specific time. Hence sequencing of transcriptome has considered as an attractive and reliable option to get a cost effective and quick interpretation about the functional element of genome in the non-model species that lack a reference genome (Garg *et al.*, 2011; Chandhini and Rejish Kumar, 2019). Study of transcriptomics helps in determination of the transcriptional structure of gene along with the initiation and terminal site (5' and 3'), splicing sites as well as the post translational modifications. In addition, it has the ability to detect the variations at expression levels of transcripts during development and under different conditions (Wang *et al.*, 2009). Prior to development of high throughput techniques, quantitative study on transcriptomics was carried out through northern blotting, nylon membrane arrays, and reverse transcriptase quantitative PCR (RT-qPCR) (Becker-Andre and Hahlbrock, 1989; Pandit *et al.*, 2018). Later discovery of modern techniques make this study more comprehensive, constructive and informative. The modern techniques related to transcriptomics may be carried out through either hybridization based method or sequencing based method (Rani and Sharma, 2017). The hybridization method i.e. microarray technology has popular due to its ability to conduct the assay of thousands transcripts simultaneously but has problem in the accuracy level for low abundance transcripts and depending upon the gene-probe interactions (Zhao *et al.*, 2014). In addition, previous knowledge about of the targeted sequences is prerequisite to construct the probe for the microarray technique and unable to detect the single nucleotide variation (Zhang *et al.*, 2014). All these biasness could be overcome by another technique RNA-Seq (Pandit *et al.*, 2018). In sequencing-based transcriptomics, the alignments of nucleotides of RNA helps the researchers to find out source DNA from which it has been derived. Though sequence-based methods i.e. SAGE, comparative EST sequencing and MPSS are more reliable and accurate tools to detect the transcript but becomes technically more challenging, relatively labor-intensive and expensive for large number of samples (Fryer *et al.*, 2002). In EST sequencing, reverse transcriptase is used to generate double-stranded complimentary DNA (cDNA) from the mRNA. These generated cDNAs are cloned to construct the libraries of transcribed genes of tissues, and the cDNA clones are randomly sequenced to obtain ESTs (Dhandapani *et al.*, 2012). But quantitative analysis is difficult due to its cost expensive and low throughput nature (Dong and Chen, 2013). Use of ESTs was reported to identify the signature genes responsible for rhizome development in ginger and turmeric and biosynthesis of secondary metabolites (curcuminoids, gingerols and terpenoids) by Koo *et al.* (2013). In addition, development of EST based marker has been reported for several spices which have been discussed in 7.2. Another technique i.e. CAGE (Cap Analysis of Gene Expression) has been developed to study sequence of the 5' capping of transcripts taking advantage of 5' capping (Shiraki *et al.*, 2003). Though the tag based sequencing method are advantageous since it provides high throughput and high level precision of gene expression but able to detect only a small portion of transcript and unable to differentiate the isoforms from each other (Brenner *et al.*, 2000; Imadi *et al.*, 2015). To overcome these problems, sequence based approach is further modified with next generation sequencing (NGS). RNA-seq, also known as whole-transcriptome shotgun sequencing, refers to the use of high-throughput sequencing technologies for characterizing the RNA content and composition of a given sample (Wolf, 2013). The rapid estimation of tissue specific expression of mRNAs at a given time by this method replacing existing methods of studying gene expression including microarrays (Egan *et al.*, 2012). RNA Seq is advantageous over microarray due to its uniqueness i.e. it covers the whole transcriptome, no prior genome sequence is needed to perform the experiment and has ability to detect all types of transcripts (novel, known transcripts, low abundant transcript) and does not show any background signal like hybridization (Anamika, 2016). Keeping view on advantages of RNA Seq over other techniques of trancriptomics, this method has been reported for several spices to fulfill several objectives. Use of RNA sequencing has been reported to study the transcripomic profile of several spices including turmeric cultivar Suvarna (Sahoo *et al.*, 2016) ; *Capsicum annuum* L. during fruit development (Martínez-López *et al.*, 2014); bell peppers (*Capsicum annuum* L.) in response to cold stress (Kong *et al.*, 2019); heat-tolerant cultivar (17CL30) and a heat-sensitive cultivar (05S180) chilli in response to heat (Wang *et al.*, 2019). Also reported to perform genome-wide transcriptional profiling of the flower buds from fertile and sterile plants of the genic male sterile–fertile line 114AB of *Capsicum annuum* (Chen *et al.*, 2015b). Among different sequencing platform for RNA Seq, the HiSeq series of sequencers from Illumina (https://www.illumina.com/systems.html) is widely applicable due to its several advantages over other platforms (Roche, PacBio, and Ion Torrent) (Anamika, 2016). Application of Illumin based sequencing have been reported to study the fruit transcriptome of black pepper (Hu *et al.*, 2015); to identify genes from wilt-sensitive ginger and wilt-resistant mango ginger confer resistance to *R. solanacearum* (Prasath *et al.*, 2014); to identify genes involved in the biosynthesis of capsaicinoids (an enzyme) in pungent pepper (Liu *et al.*, 2012).

7.5.3 Proteomics in Spices

With the help of modern techniques researchers can detect genome sequence of organism but have faced difficulty to determine the functional products from these sequences as well as their role in the multicellular physiological processes. To overcome these problems, new branch of *omics* i.e. "Proteomics" has been developed by researchers to analyze the expression and function of the gene product "proteins" (Aebersold and Cravatt, 2002). Basic study on proteomics has focused on "the products of the genes" and uses the "reverse genetics" approach i.e. starts with functionally modified protein and works back to the gene responsible for this product with respect to genomics which starts with "gene" (Sharma and Harikumar, 2013). Basic areas of proteomics cover several aspects of proteins including separation and detection of proteins as well as determination of the function and functional network of proteins (Jacobs *et al.*, 2000; Varshney and Dubey, 2009). Separation and visualization of proteins are commonly performed through conventional gel based approach. For this purpose, two-dimensional gel electrophoresis (2-DGE) has been developed and most widely reported for protein separation (in which the first dimensional separation of protein samples is carried out through isoelectric focusing method on the basis of isoelectric point of protein samples followed by second dimensional separation through sodium dodecyl sulfate polyacrylamide gel electrophoresis (SDS PAGE) on the basis of molecular weight of same samples) and finally the separated products are visualized through staining method (using either coomassie blue or silver nitrate or a fluorescent dye). Besides, 2DGE can applied to detect/characterize the post-translational modifications, mutant proteins as well as evaluate of metabolic pathways (Aslam *et al.*, 2017).

Due to these significant contributions, application of 2DGE have been reported in spices to serve several objectives of proteomics. It has been reported to perform proteomic study of anthers/buds between a CMS line along with its maintainer in *Capsicum annuum* L. (Wu *et al.*, 2013); used to identify the variation of space-induced hot pepper mutant at the protein level (Xie *et al.*, 2017); to study the variations among seed proteins of 10 inbred pepper lines (Posch *et al.*, 1992). In addition to protein separation, PAGE of 2DGE has been also reported as effective tool to study the genetic diversity plant germplasms e.g. Shamina *et al*. (1998) reported PAGE to study the genetic diversity among 15 accessions of turmeric; while SDS PAGE were reported to study the variations in soluble proteins from seeds of 29 lines of *C. annuum* (Anu and Peter, 2003); genomic relationship between 20 elite germplasms of ginger (Kar *et al*., 2014). Hence PAGE can be considered as common tool to study genetic divergence among the plant genotypes on the basis of protein content which generates fundamental principle of biochemical marker (Details has already been discussed in 7.2). Addition of silver staining in SDS-PAGE method was also reported to identify garlic and garlic products (Mochizuki *et al*., 1996). Inpite of its wide applicability, 2DGE is time consuming, non-quantitative, does not effective for hydrophobic proteins and has a limited dynamic range (Haynes and Yates, 2000). To overcome these limitations, differential imaging gel electrophoresis (DIGE) has been introduced by scientists in proteomics. In compare to 2DGE, DIGE is more accurate, reproducible, and able to quantify the protein expression differences (Ünlü *et al*., 1997; Meleady, 2018). To enhances the reproducibility and sensitivity, use of multiplexed fluorescent dyes-labeled protein is common for DIGE. 2D-DIGE commonly uses to running multiple samples (maximum 3 samples) simultaneously on a single gel to address the issue of gel-to gel variability. In this technique, different fluorescent cyanine (Cy) dyes are used for labelling proteins from different samples (Falvo *et al*., 2012; Magdeldin *et al*., 2014). Use of DIGE has been reported to investigate the chilling injury (CI) in bell pepper fruit (*Capsicum annuum*L.) (Sánchez-Bel *et al*., 2012).

After separation, the separated proteins are subjected to identification through modern tools. Most common tool to identify and characterize the protein is Mass Spectrometry (MS) (Dave *et al*., 2011). Earlier application of MS techniques was restricted due to the unavailability of suitable techniques to softly ionize the sample and had faced excessive fragmentation of sample during transfer of ionized molecules from the condensed phase into the gas phase. These problems has been eliminated with development of soft ionization techniques "Electrospray ionization (ESI)" and "Matrix-Assisted Laser Desorption/Ionization (MALDI)"(Domon and Aebersold, 2006). Use of MALDI-TOF-MS is better than ESI-MS due to its uniqueness i.e. MALDI TOF produces singly charged ions which makes data analysis more simple with respect to ESI-MS, and prior separation by chromatography does not applicable for MALDI-TOF MS analysis (Everley *et al*., 2008; Singhal *et al*., 2015). 2DGE followed by MALDI-TOF is the most widely used to study the proteome composition of various intracellular compartments in crop plants, and help to investigate quantitative protein changes upon application of various stresses or genetic perturbations (Bischof *et al.*, 2012). Use of MALDI TOF has been reported high amount of capsaicin in the placenta than that in the pericarp for *Capsicum* Fruits (Taira *et al.*, 2012). Most widely method of MS to identify proteins is a combination of peptide mass fingerprinting and amino acid sequencing via tandem MS (Finehout and Lee, 2004). 2DGE followed by tandem mass spectrometry has been reported to identify 7 differentially expressed proteins during various ripening stages of capsicum seed (Aizat *et al.*, 2013); while characterization of BGS-Haridrin from ginger was reported through SDS PAGE and ESI MALDI/MS analysis by Ramadas and Srinivas (2011). Besides mass spectrometry LC-MS/MS has been reported to identify the traces of peanut allergens in chili pepper (Vandekerckhove *et al.*, 2017). In addition to separation and identification, sequencing of protein can be done using technique such as Edman degradation method. A combined approach of proteomics i.e. isolation and separation of protein based on SDS PAGE followed by detection of mass based on mass spectrometry and N-terminal sequencing using Edman degradation principle has been reported for unique protein (G-24) of ginger (Gill, 2012); Similarly an SOD protein has been isolated and identified from the rhizome of *Z. officinale* and determined its complete amino acid sequence using N terminal sequencing, amino acid analysis, and de novo sequencing by tandem mass spectrometry (Nishiyama et al., 2017). An integrated approach of tandem mass tag labelling, high performance liquid chromatography (HPLC) fractionation, and MS based analysis has been reported to study the effect of selenium stress on the protein levels in pepper (Zhang *et al.*, 2019). HPLC-MS/MS method have also been developed the synthetic dyes (nine different) in various spices (Botek *et al.*, 2007). Use of 2D-DIGE and liquid chromatography–tandem mass spectrometry (LC–MS/MS) were used to determine differentially expressed proteins associated with cytoplasmic male sterility in pepper (*Capsicum annuum* L.) (Zhang *et al.*, 2015). Use of ion exchange chromatography has been reported to study the amino acid contents of two varieties of Ginger (Ajayi *et al.*, 2013). Use of automated Edman degradation method has been reported to sequencing a serine proteinase inhibitor from *Capsicum annuum* seeds (Antcheva *et al.*, 1996). In addition, use of X ray crystallography has been reported for find out the three dimensional structure of the N-terminal His-tagged annexin 24 (Ca32) from *Capsicum annuum* (Hofmann *et al.*, 2000).

7.5.4 Metabolomics in Spices

The latest trend of "*Omics*" involves study of metabolites of biological system. Metabolites are the end product of cellular reaction hence metabolic profiling of a cell can give us an instant idea about the physiological status of cell (Sethi and Brietzke, 2016). Variations in the metabolome (entire set of metabolites of a cell at specific time) can occur very fast within time (may be within second) (Goodacre *et al.*, 2004; Tan *et al.*, 2016). Hence study of metabolites provides us an idea about the cellular systems and decoding the functions of genes. Different types of metabolomics tools are commonly used to isolate, detect and quantify these metabolites. Among these tools, High Performance Liquid Chromatography (HPLC) is commonly used to separate the metabolites while mainly Nuclear Magnetic Resonance (NMR) and Mass Spectrometry (MS) are used to identify the metabolites (Escudero *et al.*, 2017). Applications of metabolomics in spices have reported by several researchers to identify or detect the metabolites using these above mentioned tools.

A metabolomics study was reported for 6 different *Nigella* species by Farag *et al.* (2014) using Gas Chromatography – Mass Spectrometry (GC-MS) and UPLC-PDA-MS tools. They annotated 52 metabolites (8 saponins, 10 flavonoids, 6 phenolics, 10 alkaloids, and 18 fatty acids) and found kaempferol 3-O-[glucopyranosyl- (1→2)-galactopyranosyl-(1→2)-glucopyranoside as a potential taxonomic marker to differentiate *N. sativa* from its allied species. Metabolomics study of seeds of *Cuminum cyminum* L. under saline stress has been reported by Pandey *et al.* (2015). They used Liquid Chromatography–TOF–Mass Spectrometry (LC-TOF-MS) to study the total metabolites and found 45 differentially expressed metabolites in saline stress induced in cumin plants. Metabolomic profiling of was reported by Jang *et al.* (2015) for hot pepper (*Capsicum annuum* "CM334"). They noticed higher amount of glycosides of luteolin, apigenin, and quercetin, shikimic acid, γ-aminobutyric acid (GABA), and putrescine at early stage of fruit development. A comparative metabolite profiling study has been carried out by Barbosa *et al.* (2017) for common and rare gingers. They used rhizome and leaves as metabolite source and used for this purpose. Use of Electrospray Quadrupole-Time-of-Flight Mass Spectrometry (UHPLC-ESI/QTOF) has been reported by Senizza *et al.* (2019) to differentiate between adulterated and authentic saffron, as well as to trace its geographical origin. An curcuminoid profiling has been developed using Liquid Chromatography–Quadrupole Time of Flight Mass Spectrometry (LC–QTOF-MS/MS) and Liquid Chromatography–Quadrupole Linear Ion Trap Mass Spectrometry (LC–QTRAP-MS/MS) and applied successfully to the test the quality of raw turmeric materials collected from various parts of in China and Myanmar (Jin *et al.*, 2017). Another study on metabolomics involved construction of the metabolomic profiles of four different *Curcuma* species at three developmental stages using NMR (Awin *et al.*, 2019). In addition to HPLC, use of several spectrometry techniques (UV, FT-IR, 1H NMR) were applied to construct a metabolic fingerprinting of 30 different sample of turmeric to carry out the qualitative analysis (Gad and Bouzabata, 2017). Metabolic fingerprinting of saffron has been developed through LC/MS by Guijarro-Díez *et al.* (2015). They identified six metabolites related to metabolism of kaempferol and geranic acid as glycosides using a combined approach of liquid chromatography and electrospray ionization time-of-flight mass spectrometry. GC-MS based metabolomics has been reported to determine the hexane soluble metabolites in pericarp and seeds of nine different accessions of chili fruits during the different stages of ripening (Debnath and Das, 2020). Use of HPLC reported to quantify several metabolites (i.e. crocins, picrocrocin, and safranal) of saffron grown in various environmental condition of Morocco (Lage and Cantrell, 2009). Tissue specific secondary metabolites from *Curcuma longa* L. rhizome of Indian and Chinese origin has been reported by Jaiswal *et al.* (2014). They used Ultrahigh-Performance Liquid Chromatography-Quadrupole Time-of-Flight Mass Spectrometry (UHPL/Q-TOF-MS) to detect the curcumin content in sample. Use of Gas Chromatography-Time of Flight Mass Spectrometry (GC/TOF MS) and Ultrahigh-Performance Liquid Chromatography–Quadrupole Time-of-Flight Mass Spectrometry (UPLC/Q-TOF MS) have been reported to construct the metabolite profile of *Curcuma aromatica* and *Curcuma longa* by Lee *et al.* (2014). The use of Ultra-Performance Liquid Chromatography-Quadrupole-Time-Of-Flight Mass Spectrometry (UPLC-Q-TOF-MS) and Gas Chromatography Time-Of-Flight Mass Spectrometry (GC-TOF-MS) has been reported to carry out the metabolomics study of two different *Zingiber* species (Han *et al.*, 2015). In this study, higher amount of various metabolites i.e. flavonoids, amino acids, organic acids, fatty acids etc were obtained from the parts than the root parts of plant sample. In comparative assessment of different plant parts, maximum amount of 6-gingerol was reported in ethanol extracted ginger leaves (4.9 mg/g) followed by ethanol extracted ginger flowers (2.87 mg/g) and ethanol extracted ginger rhizome (1.03 mg/g) using the HPLC (Tanweer *et al.*, 2020). In another study between normal ginger (from Ethiopia and Korea) and black gingers (from Korea and Thailand) has revealed presence of gingerol-related phenolic acid only in normal gingers with respect to methoxyflavones in black gingers (Asamenew *et al.*, 2019). They used Ultra Performance Liquid Chromatography Coupled with Diode Array Detector, Quadrupole Time-of-Flight Mass Spectrometry (UPLC–DAD–QToF–MS) to estimate these phytochemicals. Use of GC-MS and LC-MS has been reported to study the metabolite profiling during different phases of *Capsicum annuum* L. cv. Aries (Aizat *et al.*, 2014). Metabolomic study on sixty accessions of four different *Capsicum species* from the Embrapa Clima Temperado AGC was carried out by Aranha *et al.* (2017) through GC-MS approach. Metabolomic profiling of leaf tissues extracts of Mentha using GC-MS has been reported by Riahi *et al.* (2018). They identified 50 different metabolites from leaf sample of metha. Variations in metabolite profiling of three different parts of *Crocus sativus* L. has been reported with the help of using UPLC-Q-TOF-MS and 54 compounds were identified in tepals, stigmas and stamens (Xu *et al*., 2019). Use of HPLTC and GS also reported to study the photochemical of *in vitro* grown *Curcuma angustifolia* has been reported by Jena *et al*. (2018).

8

Global Breeding Research on Important Spice Crops

8.1 Black Pepper Breeding

Black pepper (*Piper nigrum* L., family:Piperaceae), known as the '**King of Spices**' and "**Black gold**", is one of the important spices since time immemorial and perhaps the oldest spice regarded globally. It is a perennial climber grown for its fruits that are used as spice with plethora of medicinal properties. Black pepper is also very important in traditional medicine (Ravindran, 2000). However, it is mainly valued for its characteristic pungency and flavour, as an ingredient in food preparations and also as a condiment.

Genetic Resources

World's largest collection of black pepper germplasm is maintained by the Indian Institute of Spices Research (IISR) at Kozhikode, Kerala, containing local cultivars, wild forms collected from the area of origin and related species. At present there are about 3466 accessions, consisting of 1266 wild relatives, 2062 cultivars, in addition to 9 exotic collections maintained in India. These collections were characterized and evaluated to estimate the genetic diversity for various yield and quality characters. Cultivar diversity is one of the principal components of diversity in black pepper. The cultivars are evolved directly from the wild *P. nigrum*.

Collection and Conservation of genetic resources: Wild populations of *Piper nigrum* are found extensively distributed in many forest areas including Western Ghats in India. It has been indicated that *P. nigrum*, a tetraploid, might have originated through hybridization between species occurring in Western Ghats. Based on morphological and biosystematic studies, Ravindran (1991) proposed three species viz., *P. wightii, P. galeatum* and *P. trichostachyon* as the presumed parents of *P. nigrum*. The fruits of all the three have small amount of pungency and flavour.

In the perspective of rapid gene erosion due to an assortment of biotic, abiotic, socio-political and economic factors, conservation of genetic resources is extremely imperative. The germplasm collections conserved in clonal field repositories are under constant risk from an assortment of biotic and abiotic stresses. Moreover, pepper being heterozygous in nature and commercially grown through vegetative propagation, conservation of germplasm in seed gene banks is mostly impractical. Hence, *in vitro* conservation becomes important as a secure substitute. Protocols for slow growth *in vitro* conservation of pepper and its related species viz. *P. barberi*, *P. colubrinum*, *P. betle* and *P. longum* were standardized (Geetha *et al.,* 1995; Nirmal Babu *et al.,* 1996b, 1999). This is achieved by maintaining cultures at reduced temperatures in the presence of osmotic inhibitors at reduced nutrient levels and by minimizing evaporation loss by using closed containers.

Long-term conservation by cryopreservation of black pepper seed has also been standardised as the seeds are recalcitrant and loses viability with the decrease in moisture content. Seeds desiccated to 12 per cent and 6 per cent moisture contents could be stored successfully by cryopreservation in liquid nitrogen (-196°C) with 45% and 10.5% survival, respectively (Chaudhury and Chandel, 1994). Recently technology for cryopreservation of pepper shoots was also developed using encapsulation and dehydration method.

A set of 20 *Piper nigrum* (Black Pepper) accessions were screened by Raghavan *et al*., (2010) to identify the extent of genetic diversity present at the molecular level using RAPD and SSR markers. Dendrogram constructed based on molecular polymorphism unveiled considerable amount of diversity among the varieties. Among this the SSR markers were found to be useful in discrimination and identification of the genotypes as it gave more genotypic specific bands.

Breeding objectives

- High yield per vine (above 3 kg fresh berries per vine),
- Enhanced quality (oleoresin above 10%, higher piperine, essential oil, etc)
- Adaptation to high altitude (above 1000 m),
- Resistance to diseases and pests, e.g., foot rot (*P. capsici*), Fusarium wilt (*Nectria haematococca* f. sp. *piperis*), black berry (*Cephaleuros virescens*), anthracnose (*Colletotrichum gloeosporioides*), 'pollu' beetle (*L. nigripennis*), pepper weevil (*Lophobaris piperis*), nematodes (*Meloidogyne incognita* and *R. similis*), etc.
- Resistance to drought.

Breeding Behaviour and Breeding Strategies

Black pepper is a predominantly self-pollinated (geitonogamy) and perennial vine. Various degrees of protogyny are encountered in *P. nigrum* . However, the protogyny is ineffective to prevent selfing as the pendent spike is copiously guaranteed of pollen from the upper flowers and there will be many spikes dehiscing pollen grains in a vine concurrently. The stigma are reported to be receptive up to 10 days after exertion and small quantities of pollen are found in pollen sac even 5 or more days after dehiscence .

Geitonogamic fertilization is precisely stimulated by positive geotropism, spatial arrangement of flowers, sequential ripening of the stigma and non-chronological dehiscence of anthers (DeWaard and Zeven 1969). Wind may help pollen dispersal by agitating the spikes/branches. Sasikumar *et al.* (1992) concluded that selfing (geitonogamy) with irregular outcrossing is the principal mode of pollination in cultivated bisexual black pepper varieties.

Breeding methods

The three major approaches followed in genetic improvement of black pepper are clonal selection, open-pollinated progeny selection and hybridization. Polyploidy breeding is also being attempted to increase the spectrum of variation.

Clonal Selection

Selection within the popular clones viz., Karimunda, Kuthiravally, Kottanadan, etc. has resulted in the production of superior varieties (Ravindran *et al.*, 1997; Sasikumar *et al.*, 2004). From about 216 accessions of the local cultivar *Karimunda* at IISR, two high-yielding varieties namely *Sreekara* (7.5 Kg / plant, dry) and *Subbakara* (7.9 Kg / plant, dry) are selected clonally. While *Panniyur-4* is an improved variety from the clones of cv. *Kuthiravally*, *PLD-2* is a selection from cv. Kottanandan. Panchami is a high-yielding line from an elite mother vine of cv. Aimpiriyan and Pournami is from cv. Ottaplackal. The later was reported to be tolerant to the root-knot nematode Meloidogyne incognita. IISR Thevam is a germplasm selection from the local cultivar Thevanmudy, which has shown field tolerance to Phytophthora foot rot disease coupled with high yield. Semangok Perak is an improved variety tolerant to Phytopthora foot rot disease, was developed from the introductions in Sarawak, Malaysia through selection.

Open-pollinated Progeny Selection

Selections from open-pollinated seedlings of popular varieties have also performed better. In spite of black pepper showing geitonogamous mode of pollination, there is also a certain percentage of crossing to help establish useful genotypes from open-pollinated progenies. Five promising genotypes from the third generation open pollinated progenies of the cultivars Balankotta, Cheriyakaniyakkadan and Kalluvally were identified by Sim (1993), which were introduced from India to Malaysia. Three black pepper varieties, Panniyur 2, Panniyur 5 and IISR Sakthi are released, so far, through open-pollinated progeny selection.

Naik *et al.*, (2013) evaluated the selected land races of black pepper (*Piper nigrum* L.) in arecanut mixed system of cultivation for high yield and disease tolerance. The variety Panniyur-1 recorded the maximum dry berry yield (2.58 kg vine^{-1}) and significantly differed among some of the accessions. Yield of Sirsi-1 (Accn. 91), Ademane (Accn. 53) and Kudragutta (Accn. 106) (2.24 kg vine^{-1}, 2.21 kg vine^{-1}, and 2.05 kg vine^{-1}, respectively) was at par with the variety Panniyur-1. However, Panniyur-1 was more susceptible to the Phytophthora wilt disease with 38 per cent disease incidence (PDI) and Ademane pepper the least with 4 PDI values.

Hybridization

It involves three major steps namely, selection of parents, developing progenies from the parents and selection of superior varieties from the progenies and their clonal multiplication. Many intervarietal hybrids were produced, involving popular varieties as parents after being evaluated for yield and quality attributes. So far, four black pepper hybrids viz., Panniyur-1, Panniyur-3, IISR Girimunda and IISR Malabar Excel have been released for cultivation.

Sasikumar *et al.*(2004) reported the first successful production of interspecific hybrids from the crosses of *P. nigrum* var. *attenuatum* and *P. nigrum* var. *barberi* with distinct anatomical and morphological features having chromosome number 2n=26. A partially fertile interspecific hybrid resistant to Phytophthora foot rot has also been evolvedfrom the crosses between *P. Nigrum* and *P. Colubrinum* (Vanaja *et al.*, 2008).

Polyploidy Breeding

Black pepper is normally diploid (2n = 52). By treating fresh seeds of Panniyur^{-1} with 0.05% colchicines, an induced tetraploid (2n = 4x = 104) Panniyur-1 was developed at IISR (Nair and Ravindran, 1992). A triploid cultivar of black pepper (2n = 78) characterized by very bold berries, loose setting and large leaves. Progenies of this triploid exhibit wide variation for somatic chromosome number and morphology (Nair *et al.*, 1993).

Mutation breeding: Attempts were made using gamma rays as source of irradiation. Apart from seeds, rooted cuttings were also irradiated and raised. Irulappan *et al.* (1982) and Ravindran *et al.* (1986) used 1-4 kr gamma rays for inducing variability in Karimunda, Panniyur 1, Kuthiravally, Kalluvally (Pulpally), Kalluvally (Malabar), Thommankodi and Aimpiriyan. Irradiation adversely affected the germination of seeds. As the dose increased, germination was delayed. The M_1 population expressed morphological abnormalities such as chlorophyll changes, twinning of seedlings and rosette leaves. Chandy *et al.* (1980) did not observe any mutants in the M_1 by treating vegetative buds with EMS and further generations were not studied. Programmes using ionizing radiations for generation of variability for selection of improved genotypes are in progress in Sri Lanka and Malaysia with some encouraging results.

Biotechnological Approaches

Micropropagation

In vitro propagation of black pepper has been reported using shoot tips, nodal segments and apical meristems. Multiple shoots can be induced using BA in the culture medium either alone or in combination with auxins. *In vitro* developped shoots could be easily rooted using growth-regulator-free basal medium. Micropropagation coupled with virus indexing and genetic fidelity testing is a good technology for obtaining disease and virus-free planting materials. Protocols for plant regeneration were standardized and plants were regeneratedfrom shoot- and leaf-derived callus cultures. The plants could be established in the field. Micropropagation through direct somatic embryogenesis and subsequent cyclic secondary somatic embryogenesis was also developed (Nair and Dutta Gupta, 2003,2006).

Protocols were standardized for micropropagation of *Piper longum, P.chaba, P. betle, P. barberi* and *P. colubrinum* (Nirmal Babu *et al*, 2005). Preliminary field performance of micropropagated plantlets of piper species indicated that they are at par with clonally propagated plants (Nirmal Babu *et al.*, 2003).

Callus and shoot regeneration were encouraged from leaf portions on Murashige and Skoog (MS) medium augmented with varied concentrations of plant growth regulators. A higher callus production (90%) was observed in explants incubated on MS medium incorporated with 1.0 mg L^{-1} 6-benzyladenine (BA) along with 0.5 mg L^{-1} gibberellic acid after 4 weeks of culture. Rooted plantlets were achieved on medium containing varied concentrations of auxins. The antioxidative enzyme activities [superoxide dismutase (SOD), peroxidase (POD), catalase (CAT), and ascorbate peroxidase (APX)] revealed that significantly higher SOD was observed in regenerated plantlets than in other tissues. However, POD, CAT, and APX were higher in callus than in other tissues. A high-performance liquid chromatography (HPLC) fingerprint analysis protocol was established for quality control in different in vitro-regenerated tissues of *P. nigrum* L. During analysis, most of the common peaks represent the active principle **"piperine"**. The chemical contents, especially piperine, showed variation from callus culture to whole plantlet regeneration (Ahmad *et al.*, 2013).

Embryogenic callus was initiated from nucellar tissue in the seeds of five cultivar varieties of Black Pepper (*Piper nigrum* L.) viz., (Karimunda, Vattamundi, Thulamundi, Arayamundi and Aimpiriyan). MS and SH media -induced somatic embryos in solid/liquid media with no hormones. Callus proliferation and somatic embryogenesis was achieved in about 8-12 weeks in MS static culture medium. Suspension cultures developed only in Karimunda and Vattamundi varieties. Better germination of the somatic embryos was obtained in $^{1}/_{2}$ strength MS medium with 30 gl^{-1} sucrose. The plants obtained were hardened and handed over to the organic cultivators for field trial (Joseph *et al.*, 2012).

Plant regeneration from callus cultures: Plant regeneration from shoot tip and leaf with or without intervening callus phase was reported by Nirmal Babu *et al.* (1997). Joseph *et al* (1996) reported cyclic somatic embryogenesis from zygotic embryos. This cyclic somatic embryogenesis from maternal tissues like integuments has tremendous potential for automated micropropagation. These systems are useful for transgenic experiments for transfer of Phytophthora resistance.

A practicable plant regeneration protocol was inexpensively recognized by Ahmad *et al.* (2011). Callogenesis was induced from petiole explants of potted plants incubated on MS media. The best callus induction 85% was observed for MS-medium supplemented with 0.5 mg/l 6 - benzyladenine (BA) after 4 weeks of culture. Subsequent transfer of callus to MS medium containing similar PGRs induced shoot regeneration. Highest shoot regeneration (92%) was recorded for 0.5 mg/l BA after 5 weeks of transfer. The best callus was produced on MS medium with 1.5 mg/l BA by shoot tip explant. Shoot regeneration was excellent on MS medium with 0.5 mg/l BA and the plantlets formed were rooted best on 1.5 mg/l IBA (Hussain *et al.,* 2011).

The organogenic potential and antioxidant potential (1, 1-diphenyl-2-picrylhydrazyl-scavenging activity) of the medicinal plant *Piper nigrum* L. (black pepper) were investigated by Ahmad *et al.*, (2010). Callus induction and shoot regeneration were induced from leaf explants of potted plants cultured on MS medium supplemented with different plant growth regulators.

The best callogenic response was observed on explants cultured for 30 days on MS medium supplemented with either 0.5 or 1.5 mg l-1 6-benzyladenine (BA)+1.0 mg l-1 alpha-naphthaleneacetic acid. An assay of the antioxidant potential of the in vitro-grown tissues revealed that the antioxidant activity of the regenerated shoots was significantly higher than that of callus and the regenerated plantlets.

A novel approach for in vitro regeneration of *Piper nigrum* L. has been applied in order to increase healthy biomass, phytochemicals and piperine production via reverse photoperiod (16hD/8hL). Leaf portions of the seed-derived plants were placed on an MS-medium fortified with different PGRs. Under 16hD/8hL, thidiazuron (TDZ; 4.0 mg L^{-1}) and BA (1.5 mg L^{-1}) was found to be the most effective (<90%) in callus induction. Healthy shoots were transferred to rooting medium and higher percentage of rooting (<90%) was observed on IBA (1.5 mg L^{-1}). These in vitro tissues were subjected to amino acid analysis, spectrophotometry, and HPLC. ARG, SER, THR, and TYR were the most abundant components out of 17 amino acids. Significant quantities of piperine content were observed in all tissues except callus cultures. These results suggest that reverse photoperiod is a promising approach for callus induction, phytochemicals and piperine production for commercial applications (Ahmad *et al.*, 2014).

The shoot apical meristem for production of plantlets from black pepper variety Sreekara was assessed by Umadevi *et al.* (2015). Meristem extension was achieved in liquid medium containing Murashige and Skoog with 0.1 mgl^{-1} kinetin and 0.5 mgl^{-1} GA_3, subsequent direct shoot induction in ½ MS with 3 mgl^{-1} BA and 1 mgl^{-1} IAA followed by shoot growth and development in ½ MS +0.5 mgl^{-1}Indole butyric acid. *In vitro* multiplication from the meristem derived plants shows the possibility for mass multiplication. The protocol will be of immense importance in rapid mass multiplication of elite germplasm, as well as for conservation of this important species.

Somatic embryogenesis and meristem-tip culture alone and in combination with the antiviral agent ribavirin were attempted to eliminate Piper yellow mottle virus (PYMoV) from infected black pepper explants. Cyclic somatic embryos obtained from the micropylar region of matured seeds collected from PYMoV-infected black pepper plants of six varieties were regenerated and hardened in the greenhouse. Testing of somatic embryo-derived plants by PCR showed virus elimination in 55–100% of the plants of the different varieties of black pepper. PYMoV elimination was increased when cyclic somatic embryos were pre-treated with ribavirin before regeneration. A protocol for meristem-tip culture of black pepper plants was developed that consisted of excising the meristem from PYMoV-infected black pepper plants and inoculating it in the regeneration medium containing antibiotics (to remove endophytic bacterial contamination), followed by rooting and hardening of the plants (Sasi and Bhat, 2018).

An efficient protocol for rapid multiplication of black pepper (*Piper nigrum* L., Piperaceae) was developed by Maju and Soniya (2012). In vitro germinated seedlings were used as the source of contamination free explants for the study. Direct induction of shoots was obtained from bulged portion of shoot tip and nodal segments cultured on Schenk and Hildebrand (SH) and Murashige and Skoog (MS) media containing various combinations of cytokinins (0.5-4 mg l^{-1} BA and 0.05-0.5 mg l^{-1} TDZ) and auxins (0.2-1 mg l^{-1} IAA, 0.05-0.5 mg l-1 NAA, 0.5-6 mg l^{-1} Kin, 0.2-0.6 mg l^{-1} 2,4-D). The best shoot proliferation (18+or-1.7 shoots/explant) response was observed in SH medium with a combination of 2.5 mg l^{-1} BA and 0.5 mg l^{-1} IAA. The resulting shoots were subsequently rooted on half strength SH medium containing 0.5 mg l^{-1} IBA within four weeks with 90% success. PCR and GUS histochemical analysis confirmed the genetic transformation.

A protocol was developed by Varghese and Bhat (2011) for an efficient Agrobactertium-mediated transformation of black pepper plants through somatic embryogenesis. Genetic fidelity test of embryogenic mass-derived plantlets by RAPD using 23 random primers revealed no genetic variation among the progenies and the parent plant. Among the antibiotics used for selection of transformants, cefotaxime at 100 micro g mL^{-1} was found to be optimum to control Agrobacterium besides its ability to promote somatic embryo proliferation. Results demonstrated for the first time an efficient transformation and regeneration of black pepper without the use of growth regulators. This simple efficient procedure would allow transformation of black pepper with genes of desirable characters.

Somaclonal variation and in vitro selection for Phytophthora foot rot tolerance: Attempts on induction of variability on somaclones for tolerance to Phytophthora foot rot resistance by Shylaja *et al.* (1994) and Nazeem *et al.* (1997) resulted in identification of tolerant somaclones through in vitro selection of calli as well as somaclones using crude culture filtrate and toxic metabolite isolated from Phytophthora capsici.

Synthetic seeds: Artificial or 'synthetic seeds' is ideal for low cost plant movement, propagation, conservation and exchange of germplasm. Synthetic seeds are developed by encapsulating in vitro developed small shoot buds in 3% calcium alginate in black pepper. These Synthetic seeds could be stored up to 9 months in sterile water with over 80 % viability (Sajina *et al.,* 1997).

Protoplast culture and development of protoclones: The 'protoplast' is a naked cell which is suitable for a variety of manipulations that are not normally possible with intact cells and hence protoplast is an important tool for parasexual modification of genetic content of cells (Vasil and Vasil, 1980). Shaji *et al.* (2005) reported high-frequency isolation of viable protoplasts from in-vitro-derived leaves of *P. nigrum* and *P. colubrinum*. Regeneration and development of protoclones were observed only in *P. colubrinum* protoplast cultures.Successful isolation and culture of protoplasts were reported in *P.nigrum* (Sim *et al.,* 1995).

Molecular characterization and development of mapping population: In recent times there is increased emphasis in molecular markers for characterization of the genotypes, genetic fingerprinting, identification and cloning of important genes, marker assisted selection and in understanding of inter relationships at molecular level. Molecular markers like RAPD, AFLP and ISSR polymorphism was used for assessment of genetic variability in black pepper and to develop finger prints for studying the inter relationships (Pradeep Kumar *et al.,* 2001, 2003, Babu *et al.,* 2003, Ganga *et al.,* 2004, Nazeem *et al.,* 2005, Keshavachandran *et al.,* 2005, Sreedevi *et al.,* 2005). *Piper nigrum* in the wild is cross pollinated and hence, the progenies are expected to carry high levels of heterozygosity (Pradeepkumar *et al* 2003).

A mapping population was developed for preparation of genetic map of black pepper (Nirmal Babu *et al.,* 2003). Johnson *et al* (2005) used male parent-specific RAPD markers for identification of hybrids in black pepper (*Piper nigrum* L.). Ajith *et al* (1997) used RAPD markers to estimate genetic fidelity of micro-propagated *Piper longum*. Banerjee *et al.* (1999) reported on male sex associated RAPD markers in *Piper longum*. Anjali *et al* (2004) studied genetic diversity amongst landraces of a dioecious and vegetatively propagated *Piper betle*– betel-vine using molecular markers. One putative RAPD marker was found to be associated with Phytophthora resistance in black pepper and the marker was converted in to SCAR.

Mani and Manjula (2010) reported the cloning and sequence characterization of two isoforms of osmotin, an antifungal PR-5 gene homologue, from a salicylic acid-induced subtracted cDNA library earlier generated in *Piper colubrinum*. The larger form of the gene is 693 bp long, encoding a 21.5 kDa protein. The smaller form comprises a 543 bp long coding sequence which code for a protein of 16.4 kDa. A notable feature of the smaller form was a prominent internal deletion of 150 bp besides certain point mutations. Cloned isoforms of osmotin from resistant species could be candidates for molecular breeding for the improvement of black pepper as well as candidates for the study of structure-based mechanism of antifungal activity attributed to PR-5 family.

Isolation of R gene candidates: Preliminary work on isolation of genes responsible for agronomically important characters, especially for biotic and a biotic stress was done. A few putative genomic and cDNA fragments associated with resistance related genes are isolated. Research work on isolation, cloning and validation of full-length genes is in progress (IISR 2005). Johnson *et al* (2005) reported a method for isolation and reverse transcription of high-quality RNA from Piper species.

In an attempt to isolate resistance genes in black pepper, molecular cloning of a cDNA fragment encoding the defense related protein β-1,3-glucanase in black pepper (*P. nigrum* L.) and methyl glutaryl CoA reductase in *Piper colubrinum* was reported (Girija, *et al.,* 2005a, b).

Bhat *et al.* (2005) reported isolation and sequencing of CMV coat protein gene infecting black pepper. Bioprospecting of novel genes from spices is attempted and the presence of pea lectin genes and tomato protease inhibitor genes was identified using heterologous probes in black pepper.

Future scope of research in breeding

With the knowledge gained so far, the future breeding strategies in black pepper must focus on convergent breeding to bring together a range of yield and quality attributes scattered in different cultivars through well planned crossing programme. This can be supplemented by marker-based selection to reduce breeding time. With respect to disease and pest resistance mobilization of genes from related taxa and species is a good option involving identification, isolation and genetic transformation if necessary. Spices being what they are form a gold mine for mining of important genes responsible for various industrial, pharmaceutical and medical processes. Industrial production of useful secondary metabolites, flavour and colouring compounds is another area of interest where the compounds are high value in nature.

8.2 Small Cardamom Breeding

Small cardamom (*Elettaria cardamomum* Maton), the Queen of Spices, enjoys a unique position in the international spices market, as one of the most sought after spices. In India, it is cultivated in the states of Kerala, Karnataka and Tamil Nadu. The natural habitat of cardamom is the evergreen forests of Western Ghats. It is found to grow within an altitude ranging between 600 and 1200 meters above MSL. It is also cultivated in parts of Guatemala, Tanzania, Sri Lanka, El Salvador, Vietnam, Laos, Cambodia and Papua New Guinea. Now Guatemala is offering stiff competition to Indian cardamom in the international market.Kerala is the largest producer of small cardamom and constitutes lion share in Indian and world market. It is used as spice for

Table 8.1: Salient Breeding Achievements of Black Pepper

Name	Pedigree	Released from	Average yield dry (kg/ha)	Oleoresin (%)	Piperine (%)	E. oil (%)	Remark
Panniyur-1 (1971)	F1 of Uthirankotta x Cheriyakaniyakadan	Pepper Research Station, Panniyur, (KAU)	1242.0	11.8	5.3	3.5	Suited to all pepper growing regions. Not suited to heavily shaded areas.
Panniyur-2 (1991)	Open-pollinated progeny selection of Balankotta	Pepper Research Station, Panniyur, (KAU)	2570.0	10.9	6.6	3.4.	Reported to be shade tolerant
Panniyur-3 (1991)	F1 of Uthirankottax Cheriyakaniyakadan	Pepper Research Station, Panniyur, (KAU)	1953.0	12.7	5.2	3.1	Late maturing. Suited to all pepper growing regions.
Panniyur-4 (1991)	Clonal selection from Kuthiravally	Pepper Research Station, Panniyur, (KAU)	1277.0	9.2	4.4	2.1	Stable yielder.
Panniyur-5 (1996)	Open-pollinated progeny selection of Perumkodi	Pepper Research Station, Panniyur, (KAU)	1098.0	12.3	5.5	3.8	Tolerant to nursery diseases and shade.
Panniyur-6 (2000)	Clonal selection of Karimunda	Pepper Research Station, Panniyur, (KAU)	2127.0	8.3	4.9	1.3	Suited to all pepper tracts.
Panniyur-7 (2000)	Open-pollinated progeny of Kalluvally	Pepper Research Station, Panniyur, (KAU)	1410.0	10.6	5.6	1.5	Suited to all pepper tracts.
Panniyur-8 (2013)	Hybrid (HB 20052) Panniyur-6 X Panniyur-5	Pepper Research Station, Panniyur, (KAU)	1365.0	12.2	5.7	1.2	Tolerant to Phytophthora foot rot, suitable for open or partial shade, regular bearer
Panniyur-9 (2018)	Open-pollinated progeny selection	Pepper Research Station, Panniyur, (KAU)	3150	12.7	6.1	5.0	Long spikes, medium berries, tolerant to quick wilt and drought, suitable for open condition
Subhakara (1990)	Clonal selection from Karimunda	Indian Institute of Spices Research, Kozhikode, Kerala	2352.0	12.4	3.4	6.0	Suited to all pepper growing regions.

Name	Pedigree	Released from	Average yield dry (kg/ha)	Oleoresin (%)	Piperine (%)	E. oil (%)	Remark
Sreekara (1990)	Clonal selection from Karimunda	Indian Institute of Spices Research, Kozhikode, Kerala	2677.0	13.0	5.1	7.0	Suited to all pepper growing regions
Panchami (2001)	Germplasm selection	Indian Institute of Spices Research, Kozhikode, Kerala	2828.0	12.5	4.7	3.4	Late maturing type suited to all pepper growing areas
Pournami (2001)	Germplasm selection	Indian Institute of Spices Research, Kozhikode, Kerala	2333.0	13.8	4.1	3.4	Tolerant to root knot nematode(M. incognita).
PLD-2 (1996)	Clonal selection from Kottanadan	IISR, Kozhikode and Central Plantation Crops Research Institute, (CPCRI), Regional Station, Palode, Kerala.	2475.0	15.5	3.3	3.5	Suited to Trivandrum and Quilon districts of Kerala, India.
IISR Thevam	Clonal selection of Thevanmundi	IISR Kozhikode	1787.0	8.15	1.6	3.1	Tolerant to foot rot disease (durable resistance). Suited to high altitudes and plains
IISR Malabar Excel	F1 of Cholamundi x Panniyur-1	IISR Kozhikode	1065.0	13.5	2.96	3.2	Suited to high altitudes and rich in oleoresin.
IISR Girimunda	F1 of Narayakodi x Neelamundi	IISR Kozhikode	2112.0	9.65	2.2	3.4	Suited to high altitudes
IISR Sakthi	Open-pollinated progeny of Perambramundi	IISR Kozhikode	5.17**	10.2	3.3	3.7	Tolerant to foot rot disease in the juvenile phase

Source: Modified from Parthasarathy, 2008, ** yield in kg/fresh vine/ha

various food preparations, in confectionaries, making perfumes and in several ayurvedic preparations (Vijayan *et al.,* 2018).

Genetic resources, germplasm collection, characterization and conservation

A total of 621 cardamom germplasm accessions are being maintained at National Active Germplasm Site (NAGS), Appangala, Karnataka which consists of 423 accessions from Appangala; 101 accessions from Pampadumpara; 41 accessions from Mudigere and 56 from Sakaleshapura. Gathering the variable germplasm and exploiting the desirable genes of these accessions through various crop improvement techniques.

Table 8.2: Morphological variations in cardamom

Sl. No.	Character(s)	Variation(s)
1	Plant growth	Robust/moderate/dwarf
2	Leaf shape	Oblong/lanceolate/ovate
3	Pseudostem (leaf sheath)	Red pigmentation/green
4	Ligule breadth	Pubescent/puberulent/glabrousNarrow
5	Colour	Red tinted green
6	Inflorescence	Basal/both basal and terminal
7	Panicle type	Prostrate/erect/semi-erect
8	Panicle branch	Multibranched/compound panicle (distal/entire/ proximal)
9	Panicle length	Long/short panicles
10	Intermodal length	Short/long
11	Capsule/ raceme	2 to 6
12	Shape of capsule	Round/oval/elongate
13	Size of capsule	Bold/medium/small
14	Colour of capsule	Golden yellow/pale green/dark green at maturity

For studying the genetic diversity, 100 accessions of small cardamom germplasm were subjected to molecular profiling by means of 25 ISSR markers. It was found that the Indian cardamom have two genetically divergent clusters namely "Kerala cluster" and "Karnataka cluster" among the germplasm collections. Characterization of export grade cardamoms from India, Sri Lanka and Guatemala based on physical, biochemical parameters and molecular techniques revealed the superiority of Indian produce.

Six research organisations *viz.*, Indian Institute of Spices Research (ICAR-IISR), Cardamom Research Centre, Indian Cardamom Research Institute (ICRI), Cardamom Research Station, Horticultural Research Station and UPASI are at present engaged in research for improvement of cardamom. Collected germplasms should be characterized on the basis of different characters *viz.* quality and quantitative characters and related to biotic and abiotic stress resistance capability. Some of the methods of conservation of cardamom germplasm includes *in-situ* and *ex-situ* methods.

Breeding objectives

Since cardamom industry has been facing many constraints like continuous use of inferior genotypes, disease/pest susceptibility, drought and hailstorm have had impact on production. To overcome all these, improved varieties with the high yielding potential having superior capsules with wider adaptability should be evolved. Cardamom crop improvement programs are, therefore, designed to achieve:

- High yield of capsule.
- Resistance to biotic stress such as katte and kokkekandu and fungal diseases.
- Tolerance to abiotic stress like drought.
- Selection of plants with bold capsules and more number of seeds per fruits.
- Higher percentage of dry capsule recovery.
- Higher seed: husk ratio.
- Product for higher quality for the aroma and flavor.
- Developing location specific varieties suited to different agro climatic condition.

Breeding methods

Clonal selection

Selection in cardamom is based on both qualitative (viz., colour, shape, size and oil content of the capsules, nature of panicle, etc.) and quantitative characters from preliminary, comparative yield trial and multi-location trials to confirm the superiority of the selected clone. Eighteen promising cardamom selections of Malabar types were evaluated with standard controls (Mudigere-1, and Mudigere-2) at the Zonal Agricultural Research Station, Mudigere. These selections differed in their yield potential and growth characters. The dry capsule yield data revealed that D-237 (475.1 kg/ha) recorded maximum yield followed by CL-692 (415.8 kg/ha). These clones produced the highest yield compared to other clones and control cultivars. These were found to be the most promising cardamom selections (Dushyanthakumar and Narayanaswamy, 2005).

At Regional Research station, Mudigere several clones like CI-757, P-8 exhibited drought tolerance. At Indian Cardamom Research Institute, Myladumpara, varieties such as Malabar and Vazhukka were found to be more susceptible to Azhukal (C.O.: *Phytophthora meadii*) disease as compared to Mysore. The improved selections such as MSS-61, MCC-12 and MCC-40 showed relative tolerance to disease.

Hybridization

Inter-varietal hybridization was made between identified superior cultivars for deriving lines with high yield, 'katte' resistance and drought tolerance. A large number of crosses have been made to combine high yield and resistance to rhizome rot and cardamom mosaic diseases. Based on performance, heterosis and combining ability, 15 hybrid combinations are short listed for further evaluation. Plant height, total tillers, bearing tillers and yield per plant were under the influence of non-additive gene action (Prasath and Venugopal, 2001).

A 8 x 8 set of crosses including reciprocals were made between elite Malabar selections of cardamom (*Elettaria cardamomum*) namely, CCS-1 and RR-1 and six cardamom mosaic virus (katte) resistant lines (NKE-12, NKE-27, NKE-34, NK3-9, NKE-3 and NKE-19) to incorporate desirable characters in the hybrids. Crossability was highest and significant in the cross NKE-19 x NKE-34 (92%) followed by NKE-12 x NKE-19 (77%). The crosses indicate the scope for combining yield and disease resistance in these crosses (Venugopal and Padmini, 2012).

Intergeneric hybridization

In an effort to bring Katte resistance from wild relatives to cultivated cardamom, inter-generic crosses were made by Parameshwar in 1977. Cross with *Ammomum neutans* set a few fruits and in other cases no fruit formation was noticed. Compatibility barriers prevented the formation of fruits in these cross combination (Madhusoodanan *et al.,* 1994).

Mutation breeding

Mutation either spontaneous or artificially induced both in seed and vegetatively propagated crops are of scientific and commercial interest. Effort has been made to develop genotypes tolerant to cardamom mosaic (katte) virus, drought, and better quality through treatment of cardamom seeds and rhizomes with different physical mutagens such as X-rays, γ-rays and chemical mutagens viz., Nitrosomethyl Urea (NMU), Diethyl Sulphate (DES) and Ethyl Methane Sulfonate (EMS), etc. However, no desirable mutant could be identified so far.

Polyploidy breeding

Polyploids were induced in cardamom by treating the sprouting seeds with 0.5 per cent aqueous solution of Colchicine (Sudharshan, 1989) being screened against 'Katte' disease infection and water stress tolerance. The polyploidy lines exhibited increased layer of epidermal cells, thick cuticle and thicker wax coating on the leaves which are the general characters associated with drought tolerance in nature.

Polyploidy was induced in 52% of the cardamom (*Elettaria cardamomum*) seeds soaked in 0.5% colchicine for 90 minutes. The resulting tetraploids (2n = 96) differed from diploids by having thicker leaves with a thick waxy coating on the abaxial surface. Pollen grain diameter in the tetraploids was greater (120 micro m) than in the diploids (114 micro m) but pollen germination was slightly lower (33.06 and 43.65%, respectively). Chromosome behaviour during meiosis was normal with few tri- and univalents during metaphase I and few laggards during anaphase I (Sudharshan, 2011).

Micropropagation

Being cross-pollinated crop, micropropagation is ideal for generating true to type and virus free planting material from high yielding clones. Kumar *et al*., (1985) reported 24 successful conversion of immature floral buds to vegetative plantlets and inflorescences from an excellent source for reducing culture contamination especially since other sources are prone to high rate of contamination. Field evaluation of tissue cultured plants of cardamom in about 100 ha area was carried out by Spices Board and the results showed that the micropropagated plants performed at par with suckers (Kuruvilla *et al*., 2005).

Three out of 5 tissue-cultured cardamom (*Elettaria cardamomum*) selections tested for yield performance were considered as promising. One selection had higher yields than control varieties Mudigere 1 and Mudigere 2, while the other lines had yields similar to the controls (Chandrappa *et al.,* 2008).

Plant regeneration and somaclonal variation

Successful regeneration of plantlets from callus of seedling explants of cardamom was reported (Nirmal Babu *et al.,* 1997). High frequency plant regeneration from rhizome and vegetative bud-derived callus cultures was used for development of somaclonal variation and selection of useful genotypes from them. Good morphological variation was observed among the somaclones. A few clones tolerant to Katte were identified (Peter *et al*., 2001).

Anther culture

Attempts on anther and microspore culture were made by Ravindran *et al.*, (2002) and plant regeneration was obtained from anther derived callus on MS medium. Plant regeneration was obtained from anther derived callus on MS medium with 0.5 mgl-1 2,4-D, 0.1 mgl -1TDZ, 0.2% Trypton along with 25% sucrose and 5% glucose or 15% sucrose and 15% glucose.

Protoplast culture

Protoplasts could be isolated successfully from leaf mesophyll tissues, collected from in vitro grown plantlets and cell suspension cultures of cardamom with a protoplast yield of 3.5 x 105 / g of leaf tissue. These protoplasts could be successfully plated on culture media and made to develop to microcalli stage (Geetha *et al.,* 2000).

Synthetic seeds

In cardamom embryogenic calli and in vitro developed shoot buds were encapsulated in 5% calcium alginate to develop synthetic seeds which could be stored upto 9 months in MS medium with 75% survival and germination (Sajina *et al.*,1997).

8.3 Ginger Breeding

Ginger consists of 47 genera and about 1400 species. Among these, 22 genera and 178 species is endemic to India. It is a herbaceous perennial having underground branched rhizome with small scales. The inner core of the rhizome is pale yellow to bluish tinge while the outer is light yellow.

Genetic resources

Six hundred and sixty eight ginger accessions have been maintained in the field gene bank of IISR. The ginger germplasm conservatory was enriched with 27 ginger accessions, eight *Zingiber* sp. and 30 related genera collected from Nagaland, Manipur and Andaman and Nicobar Islands.

A total of 43 ginger accessions collected from different locations of North East India were genotyped using selected 30 SSR markers. Among them, eight primers produced polymorphic bands.

Ginger is an essential spice crop for dayily domestic uses like turmeric. It is asexually propagated as clones (a group of genetically identical plants propagated vegetatively from a single plant). Clonal propagation is done on specific spices that produces no seeds or produces seed poorly or produces seeds only under special conditions. These are not usually subjected to self-pollination and inbreeding, and so they are highly hetyerozygous.

Table 8.3: Salient Breeding Achievements of Small Cardamom

Sl No	Variety (Year of release)	Parentage and plant type	Institute / University	Salient features
1	Mudigere-1 (1984)	Clonal selection from Malabar type	Regional Research Station, Chikmagalur	Erect and compact plant, short panicle, pale green, oval bold capsule, suitable for high density planting, moderately tolerant to thrips, pubescent leaves. Average yield – 275 kg/ha (dry)
2	Mudigere-2 (1996)	Clonal selection from open pollinated (OP) Malabar type	Regional Research Station, Chikmagalur	Early maturing, suitable for high density planting, round/ oval and bold capsule, oil 8.0%, 1,8 cineol 45.0%, α- terpenyl acetate 38.0%
3	PV 1 (1987)	A selection from Walayar collection, a Malabar type	Cardamom Research Station, KAU, Pampadumpara	An early maturing type, short panicle, elongated slightly ribbed, light green capsule, essential oil 6.8%, 1,8 cineol 33.0%, α- terpenyl acetate 46.0%, dry recovery 19.9%
4	PV 2 (2001)	A selection from OP seedlings of PV-1, Vazhukka type	Cardamom Research Station, KAU, Pampadumpara	Early maturing, unbranched lengthy panicle, long bold capsules, high dry recovery percentage (23.8%), essential oil 6.6%, field tolerant to stem borer and thrips
5	PV 3 (2018)	Malabar type	KAU	Drought tolerant, moderately tolerant to capsule borer. Dry capsule yield 416 kg/ha
6	PV 5 (2018)	Vazhukka type	KAU	Semi-erect panicle, ovoid capsule, tolerant to thrips, dry capsule yield 594 kg/ha
7	IISR Suvasini (CCS-1)(1997)	Selection from OP progeny of CL 37 from RRS Mudigere, Malabar type	Indian Institute of Spices Research, Calicut	Early maturing, suitable for high density planting, long panicle, high % of bold parrot green capsule(89), tolerant to rhizome rot, thrips, shoot/panicle/capsule borer
8	IISR Vijetha (NKE-12) (2001)	Clonal selection from field resistant plant for katte, a Malabar type	Indian Institute of Spices Research, Calicut	Virus resistant selection with high percentage of bold capsules (77.0%) oil 7.9%, 1, 8 cineol 45.0%, α- terpenyl acetate 23.4%, dry recovery 22.0%
9	IISR Avinash (2001)	A selection from OP seedlings of CCS-1 and Malabar	Indian Institute of Spices Research, Calicut	Extended flowering period, dark green capsule and retains its colour even after processing. Tolerant to rhizome rot, shoot/ panicle/capsule borer. Average yield is 847 kg/ha (dry)
10	Njallani Green Gold	A clonal selection from Vazhukka type	Farmers selection from Idukki district of Kerala	Capsule bold over 70% of the cured cardamom above 7mm Average yield is 1600 kg/ha (dry)

Breeding objectives

- Higher yield
- Wide adaptations
- Improved quality parameters (viz., higher oil, oleoresins content, and low fibre)
- Resistance to biotic stress (major pest and diseases such as rhizome rot and shoot borer)
- Resistance to abiotic stress

In India attempts were made to develop varieties through introduction, selection, mutation and polyploidy breeding. Hybridization in ginger is not feasible due to sterility.

Introduction

Since ginger is vegetatively propagated, the development of new combinations by means of hybridization is not possible, so, introductions from other sources/ places is the best alternative. Many varieties of ginger have been introduced in India from other countries based on improved quality and higher yield. For example: Maran, B1, Rio-de-Janeiro (Brazil), etc.

Selection

Most of the improved varieties of ginger are the result of direct clonal selections from local or introduced germplasms for yield, disease resistance and quality attributes.Seven varieties were so far released in ginger. Of these, Varada is the most promising. The high yielding capacity of exotic variety Rio-de-Janeiro was proved in many locations. This variety possesses a number of quality attributes (Thomas, 1966).

In terms of quality attributes

- Rio-de-Janeiro and Maran had the highest oleoresins,
- Karakkal had the highest essential oil,
- crude fibre was least in China and Nadia and high in Kuruppampadi, Maran, Jugijan, Ernad Manjeri, Nadia, Poona, Himachal Pradesh,Tura and Arippa (Nybe *et al* 1980, Kumar *et al.* 1980).

Table 8.4: Salient Breeding Achievements of Ginger

S N.	Cultivar	Institution/ Released from	Pedigree/ parentage and Plant type	FreshAv. Yield (t/ha)	Maturity (days)	Dry Recovery (%)	Crude Fiber (%)	Oleoresin (%)	E.oil (%)	Remarks
1	Suprabha	Orissa University of Agriculture and Tech. (OUAT), Pottangi	Clonal selection From Kunduli Local	16.6	229	20.5	4.4	8.9	1.9	Plumpy rhizome, wider adaptability, suitable for both early and late sowing
2	Suruchi	OUAT, Pottangi	Clonal selection from Kunduli Local	11.6	218	23.5	3.8	10.0	2.0	Bold rhizome, suitable for rainfed and irrigated conditions
3	Suravi	OUAT, Pottangi	Induced mutant of Rudrapur Local	17.5	225	23.0	4.0	10.2	2.1	Plumpy rhizome, suitable for rainfed and irrigated conditions
4	Himgiri	Y.S. Parmar University ofHorticulture and Forestry,Solan, H.P	Clonal selection from Himachal collection	13.14	230	20.2	6.4	4.3	1.6	Best for green ginger, less susceptible to rhizome rot
5	IISR-Varada	Indian Institute of Spices Research (IISR), Calicut, (Kerala)	Selection form germplasm	22.6	200	20.7	3.29-4.5	6.73	1.8	Bold rhizome, tolerant to rhizome rot. Wide adaptability
6	IISR-Mahima	IISR,Calicut, (Kerala)	Selection form germplasm	23.2	300	23.0	3.26	4.48	1.7	Plumpy, bold rhizome with low fibre content

Contd.

S N.	Cultivar	Institution/ Released from	Pedigree/ parentage and Plant type	FreshAv. Yield (t/ha)	Maturity (days)	Dry Recovery (%)	Crude Fiber (%)	Oleoresin (%)	E.oil (%)	Remarks
7	IISR-Rejatha	IISR,Calicut, (Kerala)	Selection form germplasm	22.4	200	20.8	4.0	6.34	2.4	Plumpy, round and bold rhizome with low fibre and high oil content
8	Solan Giriganga	Dr YSPUH andF, Solan	Developed from IC-593889	-	225-230	21.01	4.47	4.69	1.45	Plumpy and bold rhizomes with high dry matter recovery and < 10% incidence of rhizome rot.
9	Chandra (2018)	KAU	-	23.5	-	22.3	3.0	5.2	1.6	Suitable for moist humus rich soils of tropics
10	Chitra (2018)	KAU	-	22.1	-	23.4	3.0	4.7	1.6	Suitable for moist humus rich soils of tropics

Mutation breeding

Ten M1V5 and 102 M1V11 mutants have been maintained at IISR. Three potential mutants were identified against *Pythium* sp. (V 0.5/2, R 0.8/1 and R 1.25/4) and three potential mutants against *Ralstonia solanacearum* (HP 0.5/2, HP 0.5/15 and M 0.5/1) were multiplied (IISR Annual Report, 2018).

Physical and chemical mutagens are employed for creating variability in the sterile vegetatively propagated plants. The induced variability, once fixed, can be maintained through vegetative propagation. Use of chemical mutagen, Ethyl methane sulphonate (EMS) resulted in reduced growth and increased cytological irregularities. Use of Gamma rays also had similar effects (Rattan,1987). Almost all the induced changes appearing in the R1 generation were in chimeric form and expressed a stunted or semi-dwarfing effect, and were inhibitory on production of rhizomes (Jayachandran and Mohanachandran, 1992).

Polyploidy breeding

The rhizome buds of cv. Mahima were submerged in different concentrations of colchicines (0.025, 0.050, 0.075, and 0.1%) for 48 h to induce polyploids at IISR. Maximum sprouting was recorded in 0.025% colchicine. All the successful plants (15) have been established for further studies (IISR Annual Report, 2018).

Ramachandran and Nair (1992) reported successful induction of stable tetraploids having 2n = 44 in ginger (cv. Maran and Mananthody) by treating the sprouts with 0.25 per cent aqueous colchicines. The polyploids were more vigorous than the diploids and flowered during the second year of induction. These autotetraploids had larger rhizomes and high yield (198.71 g/plant). However, oil content of these rhizomes was lower (2.3%) than the original diploid cultivar (2.8%).

Wei-Kun Hua *et al.,* (2011) studied on generation of tetraploid ginger plants with colchicine treatment in vitro, The adventitious buds were submerged into different concentrations of colchicine water solution for different time to induce polyploidy plants. The induced buds were identified by root-tip chromosome determination and stomatal apparatus observation. All tetraploid plants showed typical polyploidy characteristics. All of the 18 selected tetraploid lines possessed higher rhizome yield and overall productivity of volatile oil and gingerol than those of the control. Adaniya and Shirai (2000) were investigated the *In vitro* induction of tetraploid ginger (*Zingiber officinale* Roscoe) and the results revealed that the treatment of shoot tip explants on agar medium containing 2.0 mg per litre BA, 0.05 mg l per litre NAA, and 0.2% (w/v) colchicine for 8 days was the most efficient way of inducing tetraploid ginger.

Micro propagation

Clonal multiplication of ginger from vegetative buds has been reported by many workers (Nirmal Babu *et al.,* 1997; Rout *et al.,* 2001). Diseases of ginger are often spread through infected seed rhizomes. Tissue culture technique would help in the production of pathogen-free planting material of elite varieties. Field evaluation of tissue cultured plants indicated that it requires at least two crop seasons to develop rhizomes of normal size that can be used as seed rhizomes for commercial cultivation. Molecular characterization of micro propagated plants indicated genetic uniformity as well as certain polymorphism (Nirmal Babu *et al.,* 2003).

Biruk *et al.*, (2012) evaluated two ginger cultivars to assess the potential of axillary buds and shoot tips as explant sources and determination of suitable growth regulators for in vitro propagation .it was found that shoot tip explants on 2 mgl^{-1} BA and 1 mgl^{-1} kinetin was found to be better than other explant-media combinations which gave an average of 7 shoots per explant within six weeks of culture.

In vitro pollination

In nature, ginger fails to set fruit. However, by supplying required nutrients to young flowers and *in vitro* pollination could be effected and develop 'fruit' and subsequently plants could be recovered from the fruits (Nirmal Babu *et al*, 1992). *In vitro* pollination was successfully attempted by Nazeem *et al* (1996) to overcome the pre-fertilization barriers and successful seed set was obtained.

Anther culture

Plant regeneration from anther callus was reported from diploid and tetraploid ginger (Nirmal Babu, 1997). Callus formation, development of roots and rhizome like structures were reported earlier from excised ginger anthers cultured on MS medium containing 2,4-D and coconut milk .

Plant regeneration and somaclonal variation

Regeneration of plantlets through callus phase has been reported from leaf, vegetative bud, ovary and anther explants (Nirmal Babu *et al.*, 1992, 1995, 1996). This system could be used for inducing somaclonal variation as conventional breeding is hampered by lack of seed set. Field evaluation of somaclones indicated variability with regard to various agronomic characters and other yield attributes. A few promising high yielding lines with tolerance to rhizome rot were identified from the somaclones. (Nirmal Babu *et al.,* 2003).

Micro rhizomes

In vitro induction of micro rhizomes in ginger was reported by many workers (NRCS, 1991; Nirmal Babu 1997). The micro rhizome derived plants have more tillers but the plant height is smaller. They gave fresh rhizome yield ranging from 100-800 g per plant with an estimated yield of 10 kg per 3 m^2 bed. In vitro formed rhizomes are genetically more stable compared to micro propagated plants an important source of disease-free planting material ideally suited for germplasm exchange, transportation and conservation (Nirmal Babu *et al.,* 2003).

Future prospects

The breeding programmes in ginger are hampered by the absence of sexual reproduction. Development of polyploidy lines to increase pollen fertility and use of *in vitro* technology for pollination and embryo rescue will open up new possibilities in ginger breeding. Till then the only possible method is identification and selection of useful natural mutants which played a major role in development of present day cultivar diversity in ginger. This can be supplemented by biotechnological approaches like in vitro selection and mutation and development of disease resistant transgenics.

8.4 Turmeric Breeding

Turmeric is an important spice crop like ginger which is asexually propagated as clones (a group of genetically identical plants propagated vegetatively from a single plant). Clonal propagation is done on specific spices that produces no seeds or produces seed poorly or produces seeds only under special conditions. These are not usually subjected to self-pollination and inbreeding, and so they are highly hetyerozygous.

Genetic resources and characterization

One thousand four hundred and four *Curcuma* accessions are being maintained in the field gene bank at IISR. The germplasm conservatory was enriched with eleven *Curcuma longa*, six *Curcuma* sp. from Nagaland and Andaman and Nicobar Islands. One hundred and fifty turmeric accessions were characterised based on different morphological traits. A total of 12 quantitative and 10 qualitative characters were recorded for each turmeric accession (IISR Annual Rep, 2018).

Breeding objectives

- Improved yielding potential (rhizome yield)
- Enhanced quality and quantity of essential oil, higher curcumin and oleoresin content

- Biotic stress (especially disease resistant breeding)
- Abiotic stress

Breeding methods

Selection

In genetically mixed population of an asexually propagated species, a superior clone may be isolated and propagated as a variety. The varietal improvement work so far attempted is only through clonal selection. Most of the improved varieties released so far are selections from germplasm or clonal selections based on location specific comparative yield evaluation trials mainly to identify turmeric types with high yield potential, high curing percentage and high curcumin content. Most of the selections have been made from the landraces collected from various parts of the country or as a result of a mutation. Improved selections were developed mainly in *C. longa* and to a lesser extent in aromatica types and *C. amada*.

Ram Chandra *et al.* (1996) evaluated turmeric at Barapani, Meghalaya. Exploitation of genetic resources in turmeric has resulted in the identification of a few breeding stocks and promising genotypes. TC-17 (RCT-1), a selection from landraces of the Jowai area of Meghalaya, had high yield potential with good quality and very high degree of disease resistance against leaf spot diseases (*Colletotrichum capsici* and *Taphrina maculans*).

Naidu and Murthy (2013) evaluated ten turmeric selections under rainfed conditions for three crop seasons in the high altitude areas of Chintapalle in the Visakhapatnam district of Andhra Pradesh. These three selections, PTS-55, BSR-1 and PTS-11 were superior to other selections and are suitable for cultivation.Shanmugasundaram *et al.* (2001) evaluated fifteen genotypes of turmeric (*Curcuma longa* L.) from five states of India for yield and curcumin content. Among them, PTS-43 yielded 33.86 t ha^{-1} fresh rhizome, 7.17 t ha^{-1} cured rhizome, 5.36% curcumin and 21.17% recovery of cured rhizome as against 28.00 t ha^{-1}, 4.48 t ha^{-1}, 3.56% and 16.00% respectively in BSR-2, the local check and 26.69 t ha^{-1}, 5.69 t ha^{-1}, 4.84% and 21.33%, respectively in the national check (var. Roma).

Hybridization

Sexual reproduction is necessary to create genetic variability in crops which are propagated by asexual methods including turmeic. Improved populations are created as a source for the selection of new clones by crossing clones with superior characters. The parent clones being heterozygous, segregation occurs in the F1 generation, thereby making each F1 generation plant a potential source of new clone and a new variety. If in case it doesn't happen, the crosses are remade, or different crosses may be made.

Studies carried out on blossom biology and viable seed set obtained through successful hybridization have opened the way for recombination breeding programmes (Sasikumar *et al.*, 1994). Renjith *et al.*, (2001) aimed at developing a hybridization technique for turmeric either through *in vivo* or *in vitro* pollination. Fruit developed after *in vivo* pollination. But in vitro-developed seed, endosperm development was not complete. The aforesaid *in vitro* techniques need refinement with respect to medium and other requirements for the germination of in vitro-produced seeds.Studies carried out on blossom biology and viable seed set obtained through successful hybridization have opened the way for recombination breeding programmes (Sasikumar *et al.*, 1994).

Mutation breeding

Induced mutations were also tested in turmeric. In Tamil Nadu, two mutants, CO - 1 and BSR - 1, identified from Erode local, were released for large scale cultivation (Balashanmugam *et al.* 1986). BSR-2 is another induced mutant from same Erode local using x-rays (Chezhiyan and Shanmuga sundaram 2000). Suroma is another mutant selection from Tsundur after irradiation with X-rays.

Preliminary trials using Colchicine, EMS and MNG at 250, 500, 1000 PPM on cv. Mydukar resulted normal spouting in all treatments. Plant height and yields were more in colchicine treatments, number of leaves was more in EMS at 1000 ppm, but yield was poor. Devi and Chezhiyan (2007) assesed the impact of gamma rays on the frequency of viable mutants in turmeric. Viable mutation frequency was higher (2.40%) at 2.0 kR in (CL144). Six types of viable mutants, i.e. plant stature (tall and dwarf), number of tillers (more and less), maturity (early and late), yield (high and low), curcumin content (high and low) and oleoresin content (high and low), were obtained.

Priya *et al.*, (2014) treated cultured explants different doses of gamma irradiation (i.e. 5, 10, 15, 20 and 25 Gy). The best result towards development of multiple shoots and number of leaves from irradiated explants was obtained in 5 Gy (4.15 and 10.75) treated explants at 60 days after inoculation respectively. Increasing doses of gamma irradiation increased adverse effect on growth of shoot clumps.

Biotechnological approaches

For vegetatively propagated turmeric crop, tissue culture techniques have been found successful. Induction of somaclonal variation is also an important breeding tool for inducing variation in crops with narrow genetic base ,viz., turmeric.

Micropropagation

Protocols for micropropagation of turmeric are available (Sunitibala *et al.*, 2001, Rahman *et al.,* 2004). This technique could be used for production of disease-free planting material of elite plants. Salvi *et al.* (2000) reported direct regeneration of shoots from immature inflorescence cultures of turmeric.Singh *et al.,* (2013) compared plantlets treated with silver nitrate (AgNO3) with the conventional rhizome. Tissue culture plants grew vigorously and taller than conventional type after six months of propagation. The highest yield potential were observed in in vitro plants (13.96 ton/ha) as compared with the conventional rhizome (6.97 ton/ha). Salvi *et al.* (2001) also reported plant regeneration from leaf callus of turmeric and RAPD analysis of regenerated plants showed variation at DNA level. Variants with high curcumin content were isolated from tissue cultured plantlets (Nadgauda *et al.,* 1982). Kuanar *et al.,* (2014) isolated Fusarium wilt disease tolerant clones of *Curcuma longa* L. cv. Suroma using continuous *in vitro* selection technique.

Sharma *et al.*, (2012) developed a protocol for the *in vitro* multiplication of turmeric using rhizome bud as explants. Maximum per cent of aseptic cultures were obtained when 0.1% $HgCl_2$ was used for 3 minutes. The cultures kept in dark (2 weeks) responded well in comparison to those kept in light. Maximum proliferation of shoots was observed in MS medium augmented with BAP (2.5 mg l^{-1}) and NAA (0.5 mg l^{-1}) within 2 weeks and average number of shoots per explant was 5.6.

Renjith *et al.*, (2001) reported *in vitro* pollination and hybridization between two short duration types VK-70 and VK-76 and reported seed set and seed development. This reduces the breeding time and helps in recombination breeding which was so far not attempted in turmeric. He and Gang (2014) developed a reliable Agrobacterium-mediated transformation protocol. Calli derived from turmeric inflorescences were used as source tissues for transformation.

Microrhizomes

It is one of the ideal methods for the production of disease free planting materials. Micro rhizome production depends on size of the multiple shoot used. Micro rhizomes produce varied in size (0.1-2.0 g) and can be planted directly in the field. Micro rhizomes regenerated plantlets are smaller in size compared with conventional method but gave reasonably big rhizomes (Nirmal Babu, 2003).

Archana *et al.,* (2014) developed an efficient protocol for the development of microrhizome and minirhizome technology in high yielding variety of turmeric, Alleppey Supreme. It showed highest response in liquid MS medium with 80 gl^{-1} sucrose in Planton culture vessels.

Molecular characterization

Sasaki *et al.,* (2004) applied single-nucleotide polymorphism analysis of the trnK gene to the identification of *Curcuma* plants. Sasikumar *et al.* (unpublished) studied over 96 Indian cultivars and related species of turmeric using RAPD profiling for establishing their interrelationships. RAPD analyses showed good polymorphism among the 96 accessions studied. The intra species polymorphism in curcuma was high as compared to the interspecies polymorphism. (IISR, 2004).

Basudeba *et al.* (2013) studied molecular cloning of resistance related sequences from wild genotypes can result in efficient turmeric improvement by evolving more effective resistance specificities.

Table 8.5: Salient Breeding Achievements of Turmeric

Varieties	Developped from	Methods of breeding	Salient features
Suvarna	IISR, Calicut	Selection from germplasm collected from Assam	Bright orange rhizome with slender fingers, field tolerant to insect, pest and diseases, yield 17.4 t/ha, dry recovery 20.0 %, curcumin 4.3 %
Suguna	IISR, Calicut	Selection from germplasm collected from AP	Early maturing, field tolerant to rhizome rot, yield 29.3 t/ha, dry recovery 20.4 %, curcumin 4.9 %
Sudarshana	IISR, Calicut	Collected from Singhat, Manipur	Early maturing, field tolerant to rhizome rot, yield 28.8 t/ha, dry recovery 20.6 %, curcumin 5.3 %
IISR Prabha	IISR, Calicut	Open pollinated progeny selection	HYV, yield 37.0 t/ha, dry recovery 19.5 %, curcumin 6.5 %
IISR Pratibha	IISR, Calicut	Open pollinated progeny selection	HYV, yield 39.1 t/ha, dry recovery 18.5 %, curcumin 6.2 %
IISR Kedaram	IISR, Calicut	Clonal selection from germplasm	Resistant to leaf blotch, yield 34.5 t/ha, dry recovery 18.9 %, curcumin 5.5 %
IISR Alleppy Supreme	IISR, Calicut	Selection from Alleppy Finger turmeric	Resistant to leaf blotch, yield 35.4 t/ha, dry recovery 19.0 %, curcumin 5.5 %
CO-1	TNAU,Coimbatore	Mutant selection from Erode local irradiated with X-rays	Bold rhizome, orange yellow, suitable for drought affected areas, yield 30.5 t/ha, dry recovery 19.5 %, curcumin 3.2 %
BSR- 1	TNAU,Coimbatore	Clonal selection from Erode local irradiated with X-rays	Suitable for drought affected areas and problematic soils of TN, yield 30.7 t/ha, dry recovery 20.5 %, curcumin 4.2 %
BSR-2	TNAU,Coimbatore	Induced mutant from Erode local	Short duration cultivar with bigger rhizomes, yield 32.7 t/ha
Roma	HARS, OUAT, Pottangi	Clonal selection from T.Sundur	Suitable for both rainfed and irrigated regions, yield 20.7 t/ha, dry recovery 31.0 %, curcumin 6.1 %
Suroma	HARS, OUAT, Pottangi	Mutant selection from Tsundur after irradiation with X-rays	Round plumpyfinger rhizome, field tolerance to leaf blotch, yield 20.0 t/ha, dry recovery 26.0 %, curcumin 6.1 %
Ranga	HARS, OUAT, Pottangi	Clonal selection from Rajpuri local	Solid and spindle shaped mother rhizome, field tolerance to leaf blotch, yield 20.0 t/ha, dry recovery 26.0 %, curcumin 6.1 %

Contd.

Varieties	Developped from	Methods of breeding	Salient features
Rasmi	HARS, OUAT, Pottangi	Clonal selection from Rajpuri local	Solid bold rhizome, yield 32.0 t/ha, dry recovery 23.0 %, curcumin 6.4 %
Krishna	HARS, OUAT, Pottangi	Clonal selection from Tekurpeta	-
Kanthi	Kerala Agril Univ	Single Plant selrection (SPS) from Mydukur	Erect leaf with broad lamina, big motherrhizomes with medium bold fingers andcloser internodes, Medium duration, Av. fresh yield (37.65 t/ha), Curcumin content (7.18%), oleoresin8.25%, essential oil 5.15%, dry recovery20.15% with a duration of 240-270days.
Sobha	Kerala Agril Uni	Single Plant selrection (SPS) from Methala Local	Mother rhizome big with medium boldand closer internodes. Inner core ofrhizomes is dark orange like alleppey.More tertiary rhizomes. Av. fresh yield (35.88 t/ha), Dryage 19.38%,curcumin content (7.39%), oleoresin(9.65%), essential oil (4.25%) with amedium duration of 240-270 days.
NDH (Narendra Saryu)	ND University of Agriculture and Technology, Kumarganj	-	High curcumin content (5-6 %), higher number of primary fingers, higher yield (10% over the local varieties)
Rajendra Sonia	RAU, Dholi, Bihar	Selection fromlocal germplasm	Bold and plumpy rhizome grows widelyunder all north Indian conditions.Av. fresh yield (23.0 t/ha), Curcumin 8.4%, essential oil 5.0% anddry recovery 18.0% with duration of225 days.
Megha Turmeric 1	ICAR Research Complex for NEH Region, Meghalaya	Selection fromLakadong type	Suitable for North East hill and NorthWest Bengal. Bold rhizomes, Av. fresh yield (27.5 t/ha) highcurcumin content 6.8% and dry recovery16.37% with a duration of 300-315days.
Suranjana (TCP-2)	UBKV, West Bengal	Clonal selection from local types ofNorth Bengal	Suitable for open and shade conditionsole or intercrop, suitable for rainfed aswell as high rainfall areas. Curcumin5.7%, oleoresin 10.9%, essential oil4.1%, dry recovery 21.2%, duration 235days, tolerant to leaf blotch and rhizomerot. Resistant to rhizome scales andmoderately resistant to shoot borer.

Modified from Shamina *et al.*, 1998

8.5 Garlic Breeding

Breeding objectives

- Higher yield
- Enhanced Bulb quality(large and uniform cloves, non-bolting)
- Resistance to biotic stress viz., diseases (purple blotch, stem phyllium blight) and pests (thrips)
- Resistance to abiotic stresses
- Longer bulb storage life

Garlic is a spice crop which is exclusively propagated by asexual/vegetative methods where individual genotypes are propagated as clones (a group of genetically identical plants propagated vegetatively from a single plant). These are not usually subjected to self-pollination and inbreeding, and so they are highly hetyerozygous.

Clonal selection

Garlic in Indian subcontinent is such non-flowering spice crop where improvement using clonal selection from available germplasm in different regions is practiced. In an experiment by Kadam *et al.,* (2016) carried out using 13 garlic cultivars and one wild mutant variety, the later exhibited maximum number of leaves per plant, length of leaves per plant, stomatal length and breadth as compare to other varieties. The variety GG-2, GG4 and G-1 have maximum number of cloves per bulb, yield of bulb per plant and number of stomata per microscopic field, respectively. Significant variability was found for all characters in different genotypes.

Izquierdo Oviedo and Gomez (2012) presénted a clonal selection 'Criollo-9'. This clone was virus-free and rejuvenated by meristem culture technique. This genotype showed a good behavior before the plagues and diseases that affect this crop, as well as a high yield potential. Ellojita *et al.,* (2014) identified a garlic selection line CBT-As153 as resistant to *Fusarium oxysporum.* Sharma *et al.* (2016) conducted an experiment among the 131 garlic genotypes using twelve morpho-agronomic traits, carried out at SVPUAT, Meerut, during Rabi-2014. Significant genetic variation for all traits indicated sufficient variability in germplasm. The genotypes viz., PG-20, K-1, GHC-1, TG-1, CFG-3, G-50 are noticed with good export potential and can be utilized for improvement of garlic germplasm. Phenotypic coefficient of variation (PCV) was higher in comparison to the genotypic coefficient of variation (GCV) and high estimates of the GCV and PCV were observed for plant height, bulb weight per plant, leaf length, number of cloves per bulb and pseudostem height.

Mutation breeding

Germplasm enhancement and breeding is difficult for garlic (*Allium sativum* L.) as it can only be vegetatively propagated. Hence, mutation induction is still the most effective way to create new varieties. Calli from two Chinese commercial garlic varieties, Zhoumou (ZM) and Yongnian (YN), were treated with different dosages (1, 3, 5 and 7 Gy) of gamma radiation. The results showed that the two genotypes differed in their sensitivity to gamma radiation. YN was sensitive to high dosage, while ZM showed better growth at higher dosage. However, the corresponding parameters in ZM were higher for 7 Gy. The methods and mutant materials in this study could be used in future garlic breeding programs (Mostafa *et al.,* 2014).

Taner *et al.,* (2004) exposed garlic cloves to 0, 5, 10, 15, 20, 25 and 30 Gy doses of Cs137. The effective mutagen dose, calculated by regression analysis, was 4.455 Gy. Aftab and Saira Banarus (2013) induced callus in MS medium containing kinetin in combination with 2,4-D (3 mg L^{-1}). The calluses obtained were exposed to UV radiations The calluses tolerant at 16% concentrations of culture filtrates were considered to be resistant to basal rot disease (*Fusarium oxysporum* f. sp.*cepae*)

Polyploidy breeding

Dixit and Chaudhary (2014) induced duplication of the diploid garlic genome by treating garlic stem discs with 0.5% (w/v) colchicine. The tetraploids possessed thicker and darker-green leaves compared to untreated control plants and also showed pronounced differences in stomatal size. In comparison to control diploid garlic plants, autotetraploid plants exhibited an average 30.7% increase in allicin concentration.

Trifluralin, a herbicide, has been reported to induce doubling of chromosomes. Cheng ZhiHui *et al.,* (2012) tested various concentrations of trifluralin and different treatment durations. The base with stem of garlic 'Gailiang' was used as the explant to induce callus on MS medium with 3 concentrations of trifluralin.The mutation and survival rate declined as the concentration and duration of trifluralin treatments increased. Garlic callus cultured on medium containing 100 micro mol L^{-1} trifluralin for 15 days demonstrated high rates of tetraploidy. 6-BA at 3.0 mg/L and IAA at 0.1 mg L^{-1} was the most effective growth regulator.

Synthetic seeds

Metha *et al.,* (2013) regenerated somatic embryos from callus. Embryo size was maximum in MS medium containing 0.5 mg L^{-1} Kn + 0.25 mg L^{-1} 2,4-D. The induction of multiple shoots was highest in MS medium supplemented with 1 mg L^{-1} Kn. Highest rooting was recorded on MS medium supplemented with 1.0 mg L^{-1} IBA. Fully mature somatic embryo of *Allium sativum* were used for the purpose of artificial seeds formation. Best seeds were prepared by using 3% sodium alginate and 1.1% $CaCl_2$.

Tissue Culture

Resende *et al.*, (2000) evaluated the growth and production of micropropagated plants (apical meristem culture) compared with plants conventionally multiplied. Treatments consisted of two forms of multiplication (meristem culture and conventional plants), four garlic cultivars, five evaluation times (30, 60, 90, 120 and 150 days after planting) and randomized block design with four replications, in a split-plot scheme.

As a result of all characteristics related to yield (mean bulb weight, total and commercial yields) of plants multiplied "in vitro" were superior compared with conventional ones. As for tissue culture multiplication, cultivars "Gravatá" and "Gigante Roxão" were the most productive, while for conventional multiplication the cultivars "Gravatá" and "Lavínia" presented the highest production.

In vitro regeneration

Pardo *et al*., (2011) studied to establish an in vitro regeneration protocol with different auxin and cytokinin level, and the explants position of the leaf and root segments, on callus induction and shoot regeneration. Callus induction on apical root segments was significantly higher. Callus were transferred to shoot regeneration medium consisting of MS medium.Maximum shoot regeneration occurred in MS medium containing 0.01 mg L^{-1} 2.4-D and 5 mg L^{-1} of Kinetin.

Safadi and Faoury (2004) conducted a study to evaluate the ability of 25 local and introduced cultivars of garlic to tolerate salinity *in vitro*. Five culture media (M1, M2, M3, M4, and M5) containing different concentrations of NaCl and $CaCl_2$ (0 and 0, 17 and 9, 34 and 18, 51 and 27, 68 and 36 mM, respectively) were used in the study. Average length of shoots in cv. Kisswany grown on M2 was 2.7-fold higher than that of the control plants, whereas no vegetative growth was observed in cv. JN/EG grown on M5. The cultivars Kisswany and Hungary were the most tolerant cultivars to salinity where the average shoot height of plants grown on M5 was more than 50% of the control.

Table 8.6: Salient Breeding Achievements of Garlic

Sl.No.	Breeding Method	Varieties	Salient features
1	Selection	TCP 129(Uttar Rangini)	Tolerant to leaf spot and leaf blotch,5.1% curcumin content and high dryrecovery (26.51%)
2.	Mass selection	Agrifound White (G-41)	Developped by NHRDF, white, short day, maturity (130-135 days), yield (12-14 t/ ha)
		Yamuna Safed G-1)	Developped by NHRDF, white, short day, maturity (120-130 days), yield (15-17 t/ ha)
		Yamuna Safed-2 (G-50)	Developped by NHRDF, white, short day, maturity (120-130 days), yield (15-20 t/ ha)
		Yamuna Safed-3 (G-282)	Developped by NHRDF, creamy white, short day, maturity (120-130 days),yield (17-20 t/ha)
		Yamuna Safed-4 (G-323)	Developped by NHRDF, white, short day, maturity (165-175 days),yield (16-17 t/ha)
		Yamuna Safed-4 (G-189)	Developped by NHRDF, white, short day, maturity (150-160 days),yield (17-18 t/ha)
3	Clonal selection	Gujarat Garlic-2, Gujarat Garlic-3	-
4	Mutation breeding	Zhoumou (ZM)	-
5	In vitro techniques	Kisswany, Hungary	-
6	Tissue Culture	Gravatá, Gigante Roxão , Lavínia	-

8.6 Clove Breeding

The clove is one of the important tree spice crops grown in India. The clove of commerce is the fully grown but unopened aromatic dried flower bud of an evergreen tree *Syzygium aromaticum*. The term `clove' is derived from the French word `cloy' and the English word `clout', both meaning `nail'. The word may have been based on likeliness of the flower bud of the clove tree to a broad-headed nail. Clove is one of the most ancient and valuable spices of the Orient, known as far back as the 1st century BC. Clove is valued as a spice and for its essential oil.

Variability

Variability in clove has been observed in the shape of trees, bearing habits, cropping season, yield,colour, shape and dimension of clove. Tree girth and leaf area are the two important morphological traits for assessing the productivity

(Balakrishnan *et al.*,1998; Kennedy and Nageswari, 2000). At IISR,Calicut, Krishnamoorthy and Rema (1992) differentiated 35 elite clove trees based on their morphological and yield traits while Balakrishnamoorthy and Kennedy (1999) identified 12 high yielding types. Variation on morphological and yield attributes was observed among the 22 clove accessions being maintained at HRS, Pechiparai (AICRPS, 2007).

Germplasm Conservation

Surveys were conducted in the major clove growing areas of Tamil Nadu and Kerala and a total of 408 accessions are conserved in the germplasm repository of clove at Indian Institute of Spices Research, Calicut.

The distinctly diverse collection of clove includes three morphological variations viz dwarf and bushy types, accession with exceptionally bold and thick clove buds (king clove) and narrow leaved variants (Krishnamoorthy and Rema 1994), whereas a reddish purple petioled Zanzibar type from Zanzibar island and one from Sri Lanka are the only introductions at present.

Very rarely clove tree with very small buds called Iilliput clove is also observed in clove.

A dwarf gene source in clove has important breeding value in developing dwarf genotypes with compact canopy size.

Some of the important species of Syzygium conserved at IISR are *S. aromaticum, S. heynianum, S.jambos, S. jambolana, S. fruticosum, S. cuminii, S. caryophyllatum, S. lanceolatum,* and *S, zeylanicum.*

Table 8.7: Micropropagation and organogenesis

Explants used	Media composition	Morphological responds	References
Shoot tip from seedling and hypocotyl	MS+ NAA+ BAP	Multiple shoots	Rai and Jagadishchandra (1986)
Shoot tips of seedling	½ MS + 3 mg / l BAP + 0.5 mg / l NAA + 15% coconut milk	-do-	Mathew and Harihara, 1990
Shoot tips node WPM+ 0.5 mg / l IBA	WPM+ 0.5 mg / l IBA	Rooting	N.R.C.S
	WPM+ 3 mg / l BAP 1 mg / l kinetin	Multipleshoot	-do-

Table 8.8: Salient Breeding achievements of clove

There is no named variety of cloves in India. In trade, there are few varieties as follows:

- Penang : Plumpy and large in size, dark red-brown, high demand across the globe.
- Amboyan: Plumpy and large in size like Penang,popular in Indian savory recipes, toothpaste, cosmetics and pomanders, lower productivity as compared to Zanzibar.
- Zanzibar: Smaller and leaner in shape as compared to Penang, black-brown, high productivity and better adaptability.

8.7 Cinnamon Breeding

Cinnamomum verum is an evergreen tree.

Cinnamon or sweet wood or true cinnamon (Dalchini or dorchini) or Ceylon cinnamon had been prized for many centuries in the Orient and is one of the earliest known tree spices in India. Cinnamon of commerce is the dried bark of *Cinnamomum verum*. It is famous for its bark and leaves which are strongly aromatic. The bark has a sweet and agreeable taste. The bark, either as small pieces or powder, is extensively used as spice or condiment. It is aromatic, astringent, stimulant and carminative. It possesses the property of checking nausea and acts as an anti-emetic. Powdered cinnamon is a constituent in chocolate preparation in Spain.

Selection

Crop improvement programmes in cinnamon were initiated in Sri Lanka by the Ceylon Institute of Scientific and Industrial Research and the Department of Export Agriculture. Eight different types of cinnamon were recognized by growers in Sri Lanka based on leaf morphology, bark pungency, grittiness of bark, leaves, etc., (Wijesekera *et al.*, 1975; Anon., 1996).

Department of Export Agriculture of Sri Lanka had identified 19 selections after screening 210 accessions. The Indian Institute of Spices Research (IISR) at Calicut (Kerala, India) has maintained 300 accessions of cinnamon and related taxa. IISR, after evaluating 291 accessions, established five elite lines based on the quality characteristics such as bark oil, oleoresin and leaf oil. One Indian accession IN189 and one Sri Lankan accession SL63 were finally selected based on regeneration capacity, fresh bark yield, dry bark yield, leaf oil, percentage of eugenol in leaf oil and cinnamaldehyde (Krishnamoorthy *et al.*,1996)

Micropropagation and plant regeneration

Micropropagation on cinnamon Chinese cassia and camphor was reported from seedlings and mature tree explants (Mini *et al.* 1997; Nirmal Babu *et al.* 1997; Huang *et al.* 1998). Multiple shoots were induced from shoot tips and nodal segments of *Cinnamomum camphora* on Woody Plant Medium (WPM) (Huang *et al.* 1998 and from a cotyledonary node on MS medium (Azad *et al.*, 2005). Successful micropropagation of Chinese cassia was reported by Inomoto and Kitani (1989) using nodal explants from seedlings on MS medium. Micropropagation protocols for *C. camphora* were developed by Nirmal Babu *et al.* (2003).

Molecular Charaterization

Genetic identification among four cinnamon species (*Cinnamomum cassia, C. zeylanicum, C. burmannii* and *C. sieboldii*) using nucleotide sequences of chloroplast DNA was studied by Kojoma *et al.* (2002). The nucleotide variation at one site in the trnL - trnF IGS and at three sites in the trnL intron was used for the correct identification of Cinnamomum species. Furthermore, single-strand conformation polymorphism (SSCP) analysis of PCR products from the trnL-trnF IGS and the trnL intron resulted in different SSCP band patterns among *Cinnamomum cassia, C. zeylanicum* and *C. burmannii*. RAPD technology was applied for the analysis of normal and excellent types of *Cinnamomum camphora* L. and genetic interrelationship of nine Cinnamomum species (Hui and Linshui, 2003 and Joy and Maridass, 2008), respectively.

Somaclonal variation

Occurrence of somaclonal variation is a potential drawback when the propagation of an elite tree is intended, where clonal fidelity is required to maintain the advantages of desired elite genotypes (Rahman and Rajora, 2001). On the other hand, stable somaclonal variations of specific type may be advantageous for the improvement of certain traits (Antonetti and Pinon, 1993; Karp, 1995; Jain *et al.,* 1998). Somatic embryogenesis is frequently regarded as the best system for the propagation of superior genotypes mostly because both root and shoot meristems are present simultaneously in somatic embryos (Jin *et al.,* 2008). Before somatic embryogenesis is used for the above purpose, the genetic fidelity of cultures needs to be determined (Vendrame *et al.,* 1999).

Table 8.9: Salient Breeding achievements of Cinnamonm

Name of variety	Released by	Avg. yield Kg/ha (Fresh)	Salient Features
Nithyashreesuperior selection	IISRCalicut	56 kg/ha in the first 4 years	High and stable, regeneration capacity (67 shoots/year),high yield, high bark recovery (40.6%), in addition to excellent quality characters (bark oil 2.7% with a very good cinnamaldehyde content 73%, bark oleoresin 8%, leaf oil 2.8%). Its very young flushes are purple which turn green in 710 days. It is recommended for all cinnamon growing region in the country, both in the plains and in high altitudes (in open condition).
Navashree	IISR Calicut	(dry) 200	crop duration (days) 3 years,bark recovery30.6%,bark oil 2.7% , leaves oil 3.0% ,oleoresin 10.0%.
Yercaud -1	Horticultural Research Station, Yercaud,TNAU.	Barkyield of 359.75 kg quills and 3800 kg of dried leaves/ha	It comes to harvest from third year onwards and can be maintained economically for 20 years. It gives high bark recovery of 35.3 per cent. It has got 2.8 and 3 per cent volatile oil in quills and leaves respectively. It gives a highregeneration capacity of 19.2harvestable shoots. Quills are sweet andlight pungent in taste.
Konkan Tej Clonal selection	Konkan Krishi Vidyapeeth, Dapoli	Fresh bark yield 789.75 kg and 3.56 kg leaf per plant.	High oil (3.2%), cinnamaldehyde (70.23%) and eugenol (6.93%)
Sugandhini selection	Aromatic and Medicinal Plant Research Station, Odakkali, KAU	Average leaf yield is18kg/tree/ yearand average barkyield is 1.2 kg/tree/year.	It is recommended for cultivation in themidlands and highlands of Kerala bothin open and as an intercrop in coconutgardens for cinnamon leaf oil. Leaf oil yield is 295ml/tree/year or 125kg/ha andthe eugenol content is 94 per cent in leafoil.
PPI (C)-1 selection	Horticultural Research Station, Pechiparai, TNAU	It gives a fresh bark yield of 980kg/ha	Bark recovery is 34.22%, leaf oil recovery 3.3% and bark oil recovery2.9%
RRL (B) C-6 selection	Regioanl Research Laboratory, CSIR, Bhubaneswar	Bark oil yield is 250kg/ha.	Bark is of high quality with 83.0%cinnamaldehyde in the oil. Leaf oilcontains 94.0% eugenol.

8.8 Nutmeg and Mace Breeding

Scientific Name:*Myristica fragrans*

Family:Myristicaceae

Origin:Moluccas Island

It was introduced to India towards the end of 18th century and in India its grown in Kerala, Tamilnadu, and Karnataka.

Fig. 8.1: Nutmeg tree with mature fruits

Nutmeg is a seed or ground spice of several species of genus Myristica. *Myristica fragrans* is a dark leaved evergreen tree cultivated for two spices derived from its fruit (Fig 8.1). Nutmeg is obtained from its seed and mace from its seed covering. It is also a commercial source of essential oil and nutmeg butter. Guatemala and Indonesia combinedly produce 68% of the world's supply.

Crop improvement

Crop improvement is very much difficult in Nutmeg and it is mainly because of :

1. The dioecy.
2. long juvenile period.
3. The difficulty in propagation and
4. single ovule in the female flower.

Genetic resources

The genus Myristica consists of about 300 species of which fifteen have been described from India. They are *M.fragrans, M. malabarica, M. magnifica,*

M. beddomi, M. contorts, M. longifolia, M. amygdalina, M. andamanica, M. attenuata, M. gibbosa, M. glabra, M.glaucescens, M. irya, M. kingie and *M. prainii* (Nybe *et al*., 2006).

- Germplasm of Nutmeg is being conserved at the Indian Institute of Spices Research, and the present holding is about 482 accessions.
- The horticulture research station Pechiparai, TNAU and KKVP, Dapoli have a collection of 22 and 87 accessions, respectively (Krishnamoorthy *et al*. 2005).

Breeding objectives

- Early bearer.
- Development of hermaphrodite varieties.
- Easy identification of male and female trees in early stage.
- High yielding varieties: more number of fruits, fruit weight, seed and mace weight, oil recovery percentage.
- Less juvenile period.
- Large uniform and round nutmegs with light brown colour.
- Low terpene content and thick mace are desirable .
- Tolerant to biotic stresses (fruit rot).

Variability in mace colour

Breeding methods

At present the crop improvement programme is limited to selection of mother plants based on their regular and heavy bearing nature.

Selection: Selection of elite types involving biotechnological options and clonal multiplication of selected mother trees through epicotyl grafting/ *in vitro* propagation techniques, avoiding a juvenile phase may result in a breakthrough in productivity (Parthasarthy *et al*., 2010).

Table 8.10: Salient Breeding achievements of Nutmeg

Variety	Released from	Breeding type	Salient features
Konkan Sugandha	Dr.B.S.K.K.V, Dapoli (1998)	Single plant selection	This is the only hermaphrodite variety released (Tree bears both male and female flowers on the same plant). Higher fruit set. It yields The nut size is 5g with 1.2g mace. 2.63 kg dry nuts (approx. 526 fruits/tree/year).
Konkan Swad	Regional Coconut Research Station, Bhatye, Ratnagiri, Maharashtra (2003)	Single plant selection	higher fruit set percentage. It's a female plant yield- 761 fruits/tree/year
Konkan Shrimanti	Dr.B.S.K.K.V. Dapoli in 2005		Female plant ,high fruit set. Yield:1.96kg/tree-mace, 9.37kg/ tree/yearnut yield.
IISR Vishwashri	IISR Calicut,kerala, In 2001	Clonal selection from elite trees	

IISR Keralashree: (Farmers participatory breeding)

- IISR Calicut, in 2013.
- Open pollinated seedling progeny,seeds are collected from elite plant at burliar. Nilagiris, Tamilnadu.
- Very bold nut, thick and entire aril.
- It is the first Nutmeg variety developed by farmers participatory breeding.

Pullan

- Kerala Agricultural University, 2018
- Recommended for cultion throughout the state of Kerala.
- Canopy conical, erect, oval nut and mace, single nut weight-10.85 g, single mace weight-1.36 g, nut yield-22.79 kg/tree, mace yield-2.86 kg/ tree

Kochukudy

- Kerala Agricultural University, 2018
- Recommended for cultion in Thrissur district of Kerala.
- Canopy conical, spreading, round nut and mace, single nut weight-11.6 g, single mace weight-2.49 g, nut yield-20.88 kg/tree, mace yield-4.48 kg/tree

Mundanatham

- Kerala Agricultural University, 2018
- Recommended for cultion in Palakkad district of Kerala.
- Canopy pyramidal, erect, oval fruits round nut, single nut weight-12.6 g, single mace weight-2.49 g, nut yield-19.7 kg/tree, mace yield-3.89 kg/ tree

Poonnathanam

- Kerala Agricultural University, 2018
- Recommended for cultion throughout the state of Kerala.
- Canopy pyramidal, erect, round fruits, attractive red mace,, single nut weight-13.85 g, single mace weight-3.02 g, nut yield-19.39 kg/tree, mace yield-4.23 kg/tree

8.9 All Spice Breeding

Pimenta dioica (L.) Merr. (syn: *P. officinalis* Lindl., *Myrtus pimenta* L., M. *dioica* L. and *Eugenia pimenta* DC (Merrill 1947) is a polygamo-dioecious evergreen tree, the dried unripe fruits of which provide the culinary spice, pimento, of commerce. Leaves are aromatic and are used in the distillation. The berries are used for the preparation of essential oil and oleoresin. It belongs to the family Myrtaceae and is known in English as allspice or pimento. The family Myrtaceae consists of about 3000 woody species, most of which is grown in the tropics. The genus *Pimenta* Lindl. consists of about 18 species of aromatic shrubs and trees native to tropical America. The genus is closely related to *Myrtus L.* and *Eugenia L.* The commercially important *Pimenta* spp. are *Pimenta dioica* (L.) Merr. providing the spice pimento (allspice) and *P. racemosa* (Mill) Moore, bay or bayrum tree providing oil of bay.

Origin and distribution

The tree is indigenous to West Indies. Jamaica is the major producer and exporter of the spice. The trees are also found in Central America (Mexico, Honduras, Guatemala, Costa Rica and Cuba) and Allspice was introduced into West Indian Islands (Grenada, Barbados, Trinidad and Puerto Rico) and in the neighbouring Caribbean islands, although its original home is in dispute.

Genetic resources

Allspice was introduced into West Indian Islands (Grenada, Barbados, Trinidad and Puerto Rico) from its place of origin. Attempts to introduce it into countries in tropical regions namely, India, Sri Lanka, Fiji, Malaysia, Singapore and

Indonesia (Java, Sumatra) have not succeeded fully due to various reasons. In India, a few trees are available in Maharashtra, Tamil Nadu, Karnataka and Kerala. In India, the plant is reported to be grown in some gardens, especially in Bengal, Bihar and Orissa. There are a few allspice trees available in Nagarcoil, Kallar, Burliar and Horticultrual Research Station, Ambalavayal.

Germplasm conservation

In IISR, Calicut has a germplasm collection of 180 trees. Variants are rarely reported in allspice. Seedling variant types with dwarf/semi-dwarf habit and short internodes and bushy nature possessing a large number of branches are being conserved in the field germplasm repository of Indian Institute of Spices Research, Calicut, Kerala, India. The leaves were smaller (about tilthe size) when compared to that of normal leaves. The variants were multiplied clonally through approach grafting and all the clones exhibited the parental character at IISR, Calicut. This dwarf and semi-dwarf plant type in allspice with large number of branches offers great potential in utilizing them in crop improvement programmes.

Crop improvement

In NRCS, Calicut, about 135 allspice accessions are being maintained. An efficient vegetative propagation technique in allspice has to be developed. Once achieved, selection of improved clone(s) of known sex from the heterogenous population would be feasible. In addition to yield, attention has to be paid to breed tree with low branching habit to facilitate easy harvesting. Resistance to diseases, particularly leaf rot needs to be located. Flowering and fruit set in plants are very poor and is a major handicap in taking up any breeding work on this crop in plains. Drenching the plant basins with low concentration of Paclobutrazol (1 gm. a.i) indicated profuse flowering, under Peruvannamuzhi conditions.

Future vision

With a sound breeding programme on tree spices, as outlined above, it is possible to evolve improved varieties of tree spices and thereby increase the overall production of all spices in our country.

8.10 Vanilla Breeding

Vanilla germplasm are being maintained in the repository, which includes a flower colour variant collected from Andaman and Nicobar islands. Comparative anatomical analysis of different vanilla species was carried out. Interspecific hybridization was made between *Vanilla planifolia* and *V. aphylla*. Reciprocal crosses were conducted between *V. planifolia* and *V.*

tahitensis (species reported as resistant to root rot disease) and high percent of fruit set was observed in both the crosses. Fifty interspecific hybrids each of *V. planifolia* × *V. tahitensis*, *V. tahitensis* × *V. planifolia* and selfed progenies of *V. tahitensis* were established *ex vitro*. Chromosome number analysis of two interspecific hybrids between *V. planifolia* and *V. tahitensis* showed 2n=30 in one (PT-5) and 2n=32 in other (PT-17).

Protocols for micro propagation through direct shoot multiplication as well as callus regeneration were standardized. Root rot and wilting were found to be the major problems in most of the plantations. Root rot incidence ranged from 5 to 100%. Mosaic and necrosis were also observed in all the plantations and the incidence ranged from 2 to 80%. *Cucumber mosaic virus* (CMV) of vanilla was characterized on the basis of biological and coat protein (CP) nucleotide sequence properties, which showed that CMV infecting vanilla belongs to subgroup IB. A virus causing mild chlorotic mottle and streaks on leaves of vanilla was identified as a strain of *Cymbidium mosaic virus* (CymMV) based on coat protein gene sequence comparison and phylogenetic studies. Another virus associated with necrosis and mosaic on vanilla was identified as a strain of *Bean common mosaic virus* (BCMV) based on coat protein gene sequence comparison and phylogenetic studies.

In vitro screening of microbial antagonists against major pathogens

Chaetomium isolates were screened under *in vitro* conditions against two major pathogens of vanilla-*Phytophthora meadii* and *Colletotrichum gloeosporioides*. Among the isolates, FVREP4 showed maximum reduction in growth of pathogens by 56.88 and 83.73% for *P. meadii* and *C. gloeosporioides*, respectively.

Endophytic and epiphytic bacteria isolated from healthy vanilla plant parts were screened against *P. meadii* and the inhibition ranged from 0.0 to 53.0%. Among the isolates tested, VSEN7 and VAREN4 showed maximum (>50%) inhibition of the pathogen. However, none of them showed more than 50.0% inhibition of *Sclerotium rolfsii,* causing stem rot.

In vitro screening of fungicides against major pathogens of vanilla

Four fungicides *viz.*, propioconazole (45% EC), hexaconazole (5% EC), carbendazim 12% + mancozeb 63% (WP) and metalaxyl 8%+ mancozeb 64% (WP) were tested at five different concentrations (500, 1000, 1500, 2000 and 2500 ppm) against *P. meadii, S. rolfsii* and *C.gloeosporioides.* All the fungicides showed 100% inhibition of *P. Meadii* and *C. Gloeosporioides* at 500 ppm concentration while the fungicides - propioconazole (45% EC) and hexaconazole (5% EC) showed 100 % inhibition of *S. rolfsii* at 1500 ppm only (IISR Annual Rep, 2018) .

Four major types of vanilla beans are distinguished in the world:

- The Bourbon Vanilla (grown in Madagaskar, Camorous and Reunion), ranks top in quality(whole/split), dark brown to black shining Red, vanillin content:1.8-2.4 %
- The Java Vanilla (grown in island Java in Indonesia)
- The Bourbon-like Vanilla (grown in island Bali in Indonesia)
- Mexican Vanilla (grown mainly in Mexico)

8.11 Coriander Breeding

Coriander is one of the most vital seed spice crops in India. The most common breeding procedure used in coriander is selection and crossing is non-existent. This possible reason may be the herbaceous and delicate nature of the umbels, umbellets and florets thus making the crop difficult or less amenable for crossing (Diederichsen 1996). The pests and diseases such as powdery mildew, stem gall, wilt and white flies are serious in nature and need crop improvement interventions to get resistant varieties.

Breeding Objectives

- High yielding varieties
- Earliness
- Varieties with higher essential oil content
- Biotic stress tolerance
- Abiotic stress tolerance
- Varieties adaptable for different agro-climatic regions

Large variability exists in the yield and the yield contributing characters in coriander. The characters such as days to flowering, days to maturity and number of secondary branches are to be taken into account while selecting parents for hybridization. The methods of breeding that can be attempted are selection, hybridization and mutation breeding.

Selection

Early ecotypes were selected in dense culture under stress through short-cut selection. Mother plants selection was practiced in the full-sibbed population in the oscillated environment. This program includes several lines promising in the essential oil content, essential oil yield, crude fat and protein contents (%) as well as fruit yield per plant. The selected population exhibited apparent stability in different studied characters measured through stability against micro-environment, as well as flexibility in interactions with seasonal variatiion which enables maximizing fruit yield through agromechanics. In coriander, selected populations expressed high degree of association among traits and was characterized by short season of 120 days. The antimicrobial activity of the intact population was outlined.

Hybridization

Coriander is the most important seed spice crop in India. Selection is the most common breeding procedure used in coriander and crossing is non-existent. Giridhar et al. (2007) conducted a study to promote crossing as a breeding technique in coriander. It was found that coriander was protandrous and the sex ratio was 1.75 to 2.07. The stigma was receptive from the 3rd day to 6th day of anthesis indicating that artificial pollination on emasculated florets must be done repeatedly on 3rd and 4th day of anthesis. Pollen was viable for two days in the field. Fresh pollen can be stored in incubator at 25°C for three weeks with 88% retention of viability. Emasculation of florets was carried out with the help of binocular loupe. Emasculation either in the morning before anthesis or previous day evening was found to be suitable. For demonstrating the technique, four parents were reciprocally crossed. Only three crosses were successful indicating the differences in combining ability of the parents. Mean success among the crosses was 23%, fruit set among the florets varied from 9.52-83.3% depending on the cross combination.

Mutation Breeding

Mutation breeding is more advisable for this crop because the exploitation of exotic variability through recombination breeding is tedious and hybridization is difficult due to the tiny size of its flowers. Variation is the prerequisite for any improvement programme. In a study conducted with 60 genotypes of coriander, reported wide range of variability for plant height and number of umbels per plant whereas, low and significant variability was recorded for umbel lets per umbel (Suthanthirapandian *et al.*, 1980). Irradiated seeds of the coriander variety CO-1 with 5,10,15,20 and 25 KR doses and studied its effect on germination, number of primary branches, plant height number of umbels per plant and number of seeds per plant. The results, showed significant differences between treatments for all the characters except plant height and germination percentage in M1 generation.

Coriander is one of the major seed spice grown for leaf and seed purpose. with a view to estimate the nature and extent of genetic variability induced by three doses of gamma irradiation, three doses of EMS and combination of gamma rays with EMS in Coriander variety Swathi. Separately harvested M1 plants were used to raise M2 generation. Observations on twelve quantitative traits were recorded on five randomly selected normal looking plants in each family of a treatment. Of the twelve characters studied, induced variability was higher for number of primary and secondary branches, leaf yield, number of leaves, number of umbels and yield per plant suggesting scope of improvement through selection in these characters.

Research work at Indian Institute of Horticultural Research has resulted in development of a leafy coriander variety, Arka Isha, with a high yield potential and good aroma. It is a multicut variety where the plants are bushy, leaves are broad and leaf lobes are short, and the variety is late flowering. Yield is 3.74t ha^{-1} by pulling at 40 days after sowing, and 11.98 t ha^{-1} by cutting. Leaves have 167.05mg $100g^{-1}$ of Vitamin C, with good aroma and keeping-quality (Varalakshmi *et al.*, 2012).

In vitro studies

Hypocotyl segments and zygotic embryos of coriander formed embryogenically at frequencies of up to 75% when cultured on MS medium supplemented with 1 mgl^{-1} 2,4-D. Calli were transferred to MS liquid medium with 1 mgl^{-1} 2,4-D to initiate cell suspension cultures. Embryogenic cells became finely dispersible in the medium as the subculture proceeded. Cultures were transferred to a nitrogen compound enriched liquid MS medium containing 2% sucrose and 0.1 mgl^{-1} 2,4-D, and cultured two weeks before plating on MS basal medium. Approximately 75% of cell aggregates (1 to 2 mm in diameter) underwent development into globular to cotyledonary somatic embryos after two weeks of planting. Most of the embryos were subsequently regenerated into plantlets. Regenerants were successfully transplanted to potting soil and grown to maturity in a phytotron (Suk *et al.*, 1996).

Table 8.11: Salient Breeding achievements of Coriander

Sl No.	Variety	Pedigree/Parentage	Avg. yield (kg/ha)	Essential Oil %	Duration (days)	Salient features
1	Guj. Cor.1	Selection from germplasm	1100	0.35	112	Suitable for early sowing, erect plant,round bold grains, moderately tolerant to wilt and powdery mildew
2	Co.1	Selection from Koilpatti local	440	0.27	110	A variety with small statuted plant, suitable for rainfed areas and for greens and grains, small grain
3	Co.2	Reselection from culture P2 of Gujarat	520	0.40	90-100	A dual purpose variety, suitable for saline, and alkaline and drought prone areas seeds oblong,medium.
4	Co.3	Reselection from Acc.695 of IARI, New Delhi type	650	0.38-0.41	85-95	A dual purpose variety, good yielder, medium sized grains, suitable for both rainfed and irrigated condition, rabi as well as kharif season. Field tolerant to powdery mildew, wilt and grain mould.
5	Co.4	Reselection from germplasm ATP77 guntur collection	600	0.4	65-70	Early maturing variety suitable for both rainfed and irrigated condition; grains oblong and medium; field tolerant to wilt and grain mould
6	Guj.Cor.2	Reselection from Co.2	1450	0.40	110	Semi spreading type, suitable for early sowing,moderately tolerant to powdery mildew,grainsoblong,lodging and shattering resistant.
7	Rajendra Swathi (RD-44)	Pureline selection from Muzaffarpur collection	1300	0.65	100	Medium sized plant with fine,aromatic round grains,Suitable for intercropping, field tolerance to aphids
8	Sadhana	Mass selection from local Alur collection	1025	0.20	95-110	A dual purpose,semi-erect variety; Suitable for rainfed condition field tolerance to white fly, mites and aphids. A mid-late variety withstands moisture stress, responded well to input management under optimum moisture level.

Contd.

Sl No.	Variety	Pedigree/Parentage	Avg. yield (kg/ha)	Essential Oil %	Duration (days)	Salient features
9	Swathi	Mass selection from Nandyalgermplasm	855	0.30	82-85	Plants medium size semi-erect type, early maturing variety, suitable for rainfed condition, and late sown season. Field tolerant to white fly,moderately tolerant to disease. Suits well to the areas where the soil moisture retentiveness in compariably less, being early maturity. It escapes powdery mildew disease.
10	CS 287	Reselection from Guntur collection	600	-	79-97	Early maturing variety, suitable for both rainfed and irrigated condition. Field tolerant to wilt and grain mould.
11	Sindhu	Mass selection germplasm,Warangal local	1000	0.40	100-110	Oval medium breakable grains, suitable for rainfed areas, tolerant to wilt, powdery mildew as well as drought condition,medium duration.
12	HisarAnand	Mass selection from Haryana collection	1400	0.35	-	A medium tall dual purpose variety,oval medium sized seeds, wider adaptability to different soil conditions.Resistant to lodging due to spreading habit.
13	HisarSugandh	Mass selection from indigenous germplasm	1400			Suitable for irrigated conditions.Resistant to stem gall diseases.
14	HisarSurabhi	Mass selection from local germplasm	1800	0.4-0.5	130-140	Bushy erect plant type, seed medium, oblong; tolerant to frost, less susceptabile to aphids, medium duration
15	Azad Dhania-1	Mass selection from Kalyanpurgermplasm collection	1000	0.29	120-125	Erect, early branching, number of umbellates per umbel 5, tolerant to moisture stress, powdery mildew and aphids.
16	Pant Haritima	Selection from local type Pant Dhania	1200	0.4	150-160	Tall erect plant, a dual purpose type, good yielder of leaves, smaller seeds with high oil. Resistant to stem gall..

Contd.

Sl No.	Variety	Pedigree/Parentage	Avg. yield (kg/ha)	Essential Oil %	Duration (days)	Salient features
17	DWA 3*	Purelineselection from Karnataka collection	400	0.27	-	A dual purpose variety and for seed production in rabi crop, moderately tolerant to powdery mildew, black clay soils are best suited
18	CIMPOS-33	Selection from germplasm introduced from Bulgaria	2100	1.3	-	Tall erect, compact, profusely branching and flowering, grains small and bold. Mainly recommended for oil production.
19	ACR- 01-256	Reselection from EC-467683 from Russia	1100	0.35-5	-	Dual purpose variety, long duration, resistant to stem gall and wilt.
20	RCr 20	Recurrent half sib election from Jaipur local	900	0.25	100-110	Medium sized bushy plant suitable for rainfed crop or limited moisture condition and heavy soils of south Rajastan. Moderately resistant to stem gall, bold grains,early maturity.
21	RCr 41	Recurrent half sib selection from local type from "Kotta"	909	0.25	130-140	A tall erect plant with thick stem.Grows well under irrigated conditions, resistant to stem gall, wilt and moderately resistant to powdery mildew ;small seeds (9.3g/1000 seed),long duration varitey
22	RCr 435	Recurrent selection from local germplasm from Jalore	1000	0.33	110-130	Plants are bushy, erect, bold seeds, medium sized, medium maturing variety, adapted for irrigated condition moderately resistant to root knot and powdery mildew.
23	RCr 436	Recurrent half sib selection from local germplasm from Kotta	1100	0.33	90-100	Plants semi dwarf, bushy type with quick early growth and bold seeds.Resistant to root rot and root knot nematodes most suitable for limited moisture condition and heavy soils of south Rajasthan
24	RCr446	Half sib selection from local type from Jaipur local	1200	0.33	110-130	Plants tall, are leafy erect with higher number of seeds per umbel.Seeds are medium in size and moderately resistant to stem gall.

Contd.

Sl No.	Variety	Pedigree/Parentage	Avg. yield (kg/ha)	Essential Oil %	Duration (days)	Salient features
25	RCr 684	Mutation breeding of gamma rays. Induced mutant of Rcr-20	990	0.32	100-120	A variety, resistant to stemgall and less susceptible the powdery mildew .Adapted to medium heavy textured soil and sandy loam soil under irrigation. Seeds of the variety are bold.Plants are toleran and erect with higher number of seeds per umbel,medium maturity.
26	RCr 480	Recurrent selection based on individual plant progeny (half sib) performanmce in accession N o. 480	1325	0.44	110-130	Plants are bushy, erect, with higher number of seeds per umbel
27	RCr 728	Recurrent selection based on individual plant progeny (half sib) performanmce of UD-728, a local collection from Mandsaur (MP)	1370	0.38	130-140	Plants are bushy, erect, with higher number of seeds per umbel
28	Ajmer Coriander 3	ICAR-NRC on Seed Spices, Ajmer	1300	0.55	-	High volatile oil (0.55 %) with high linalool (75.42 %), high (13.09 q/ha) and stable yielder.
29	Rajendra Dhania-3	Dr RPCAU, Dholi	1400	0.52	-	Climate resilient variety with high yield (14.09 q/ha) and high oil content (0.52 %)
30	JD (SI)-1	JNKVV, Jabbalpur	1400	0.67	-	High oil type (0.67 %) with high yield potential (14.14 q/ha)
31	Chattishgarh Sri Chandrahansini Dhania-2	IGKV, Raigarh	-	-	-	Climate resilient variety suitable for both leafs and seeds, moderately resistant to powdery mildew and aphids
32	Gujarat Coriander-3	(CRSS), (SDAU), Jagudan	-	-	-	High quality, high volatile oil (0.52 %) with high linalool (72.16 %) and high yield potential (16.94 q/ha)
33	Ajmer Coriander 2	ICAR-NRC on Seed Spices, Ajmer	-	-	-	Tolerant to stem gall resistance, high linalool (71.7 %), early maturing type.

8.12 Fennel Breeding

Foeniculam vulgare Miller (syn. *Foeniculam officinale)*

The diploid chromosome number is 2n= 22

A large number of varieties and races differing in size, odour and taste of fruits exist among wild and cultivated fennel but they are hardly distinguishable externally from one another. They are regarded as races, varieties or sub species of *F. vulgare*. The Indian Fennel (*F. vulgare* Mill.) is a stout glabrous aromatic herb of 100 to 180 cm high.

Germplasm Collection, Maintenance and Evaluation

First systematic collection of the fennel germplasm from India and abroad was done at IARI, New Delhi. Evaluation of this collection led to the identification of a number of superior entries namely ND (D) 186, ND (D) 32, Np (J) 13 etc. which gave 18-40% more yield than the local check. Being a cross pollinated crop, each collection is heterozygous and heterogenous. Maintenance of collections poses a serious problem. In view of the fact that pollination in this crop is brought about by bees and insects, an isolation of at least 500 meters is required to exclude the alien pollain grains. At present, therefore, sib-mating under muslin cloth cages is practiced to maintain the germplasm.

Inbred lines in fennel can be easily developed and maintained by selfing through bagging individual umbels, as the seed set of bagged umbels is good. The number of inbreds normally required to include the whole range of variability existing in the germplasm is large, which requires tedious record keeping and their maintenance is also difficult. This approach of maintaining the germplasm variability, therefore, can be used in conjuction with other methods described above.

Conservation under in vitro

Dekkers *et al.,* (1991), Nirmal Babu *et al.,* (1994), Geetha *et al*., and Nirmal Babu *et al.,* (1996) reported maintenance of germplasm *in vitro* by slow growth. Suuspension of embryogenic cell lines of fennel, conserved up to 12 weeks produced normal plants upon transfer to normal laboratory conditions (Umestsu *et al.,* 1995).

Breeding objectives

- To develop high yielding varieties.
- To develop varities having high oil, oleoresin, curcumin and low fibers.
- To develop varities for wide adaptation.

- To develop varities resistant to biotic stress (major diseases and pests)
- To develop varities resistant to abiotic stress

Breeding methods

Improvement of locally adapted populations from different geographic areas through selection to develop suitable varieties is the major breeding work. Scope of heterosis breeding is also indicated as apotential method.

Selection

At present, mass selection and single plant selection based on the performance of progeny are used to develop varities in fennel. Improved varities like PF-35 and S-7-9 in Gujrat, Lam Sel.1 and 2 in A.P and UF (M)-1 and UF-31 in Rajasthan have been developed by mass-selection. Similarly, Rajasthan variety UF-32 was produced by single plant selection with higher yield potential owing to higher number of umbellets per plant.CO-1 variety of fennel from TNAU is a re-selection from PF-35 introduced from Gujarat CO-1.

An experiment was conducted on fennel cultivars RF-125 and UF-143 to evaluate half-sib progenies. Significant differences between families for days to 50% flowering and seed yield per plant in both the populations was observed, while harvest index and test weight were significant in RF-125 and UF-143, respectively. Narrow-sense heritability was high for plant height, umbels per plant, seeds per umbel, and biological yield in both populations. The expected gain due to selection between families was lower compared to the expected gain from selection within families for most of the characters in both the populations (Singh, *et al.*, 2003).

Sharangi, *et al,*. (2003) evaluated 7 fennel germplasm accessions (RF-11, RF-14, RF-15, RF-22, RF-29, RF-101 and RF-125). RF-11 produced the tallest plants (150.85 cm) with the highest number of primary branches (8.23) and umbels (7.20) per plant, and seeds per main umbel (230.23). RF-14 had the highest number of secondary branches (18.26) and umbels (14.33) per plant, number of seeds per primary (253.83) and secondary (140.76) umbel, seed yield per hectare (10.70 q/ha) and harvest index (28.4). The highest 1000-seed weight (6.03 g) was recorded for RF-29. The minimum number of days to 50% flowering (82.3) was observed in RF-125.

The nature and amount of genetic variability, genetic associations, co-heritability, and path coefficients were studied by Lal (2007) in eight quantitative traits of 37 genotypes of fennel. Seed yield/plot was significantly and positively correlated with oil content at both genotypic and phenotypic level. Seed yield was negatively correlated with days to flower 50%. But only diameter of main stem and oil content made the highest amount of direct and indirect contribution to t-anethole content. Seed yield/plant showed the largest heritability in broad sense, genetic advance (GA), and co-heritability with t-anethole content than the other traits. Therefore, seed yield and oil content could serve as better selection criteria to improve t-anethole content in fennel

Heterosis Breeding

Nine genetically diverse varieties of fennel were crossed in varietal diallel cross combination excluding reciprocals. Variety heterosis (hj) effects were significant for test weight. Specific heterosis (Sjj) effects were found significant for days to flowering, umbels per plant, umbellets per umbel, seeds per umbel, test weight, biological yield per plant and harvest index including seed yield per plant. Based on Vj and hj values, parents RF - 125 and JF - 29 were found superior. These two parents appeared in majority of crosses which showed superiority on the basis of Sjj, effects. The other parents worth considering are UF 134 and UF(M)-1. These varieties represented a good choice to initiate inter-population improvement. There was good agreement between Sjj, effects and heterobeltiosis. Cross UF(M)-1 x UF-134 showed significant and positive Sjj, effect for seed yield per plant, umbellets per umbel, seeds per umbel and harvest index (Dashora *et al*., 2000).

Ramanujan and Tiwari (1970) obsereved hybrid vigourto the extent of 24.6 to 108.9 % increase over the best pure lines when the hybrids were produced through crossing elite lines with a cytoplasmic male sterile female parent. Exploitation of heterosis through use of hybrid varieties, therefore, may be a successful proportion if cytoplasmic genetic male sterility is discovered and utilized in production of hybrid seeds.

Ramanujan *et al.,* (1994) advocated the use of polycross method of breeding as an alternative method to hybrid varieties, to exploit hybrid vigor till suitable form of male sterility is discovered. The produce of poly cross-nursery may be successfully used to develop improved varieties through recurrent selection.

Singh (2000) in a half varietal diallel cross among nine varieties of fennel observed significant differences among varieties and their crosses for almost all the characters studied including seed yield. Season x entry sum of squares was also significant for all the characters except days to 50% flowering and harvest index. The heterotic component accounted for more than 70% of the mean sum of squares due to genotypes indicating the importance of heterosis in the inheritance.

Mutation Breeding

Das *et al.,* (1989) reported that treatment of *C. sativum* and *F. vulgare* with 3 krad gamma radiation produced a dwarf, bushy, late-flowering coriander mutant and a tall early flowering fennel mutant in the M3. Analysis of the seed proteins of these mutants showed that the coriander mutant had a lower protein content, fewer low MW protein bands and new high MW protein bands. The fennel mutant had a higher seed protein content and new high MW protein bands.

Sixteen (15 viable types) macromutant (chloroxantha, thick stem I, thick stem II, slender stem, pigmented stem, lax branching I, lax branching II, funnel, bushy, drooping branched, dwarf, broad pinnae, elongated pinnae, narrow pinnae, early flowering and late flowering) types have been induced in the M2 generation of celery (*A. graveolens*), fennel (*F. vulgare*) and ajowan (*T. ammi*) following treatments with gamma-rays (2, 4, 8 and 10 kR) and EMS (0-25% and 0.50%-2 h and 4 h durations). Total mutation frequency was noted to be higher in celery (15.5%) than fennel (4.3%) and ajowan (4.2%). Gamma irradiation was more potent in inducing higher frequency of mutation than EMS in celery and ajowan (Paul, *et al.,* 2005).

Biotechnological Approaches in Fennel

As the crop is valued for quality of the produce, it is essential to identify genotypes with high quality attributes and use such genotypes for further crop improvement programmes. In this context, biotechnological tools can be utilized to multiply the elite genotypes as well as for crop improvement prorramme and for conservation of genetic resources irrespective of climatic conditions.

Callus cultures were produced from cotyledons or hypocotyls of *A. acutiloba* and *F. vulgare* in media with 0.5, 1 or 2 mg 2,4-D and 0, 0.5 or 1 mg kinetin/litre (Ohga *et al.,* 1989). Embryogenesis took place with 2,4-D and kinetin in *A. acutiloba* and with only 2,4-D in *F. vulgare*, both with 3% sucrose in darkness. Plant regeneration from embryoids took place in 16-h days with no plant hormones and 2% sucrose. Embryogenesis took 2.5-3 months and plant regeneration 3-3.5 months.

Somatic embryos were cultured by Song *et al*, (1991) on MS medium supplemented with various concentrations and combinations of growth regulators. Regeneration of normal plantlets was higher in hormone-free medium (77%) than in media containing 2,4-D (63%), benzyladenine (BA; 67%) or kinetin (47%) alone. The best combination of growth regulators was 0.01 mg BA + 0.01 mg 2,4-D/litre (63%). Gibberellic acid alone or in combination gave lower regeneration frequencies.

Auxillary and apical buds from 11 Zefa Fino and 4 Zefa Tardo plants were cultured on modified MS media. All 15 genotypes produced shoots. Some shoots were rooted and some used for somatic embryogenesis induction. Somatic embryo formation was genotype dependent, occurring only in 1 Zefa Fino and 1 Zefa Tardo genotype, as well as tissue dependent, being particularly good with nodal segment explants. Three-hundred SE plants were transferred to soil, but only 10% acclimatized to field conditions (Theiler *et al.,* 1992).

The effect of gibberellic acid (GA3) on somatic embryogenesis from petiole fragments excised from micropropagated fennel plantlets were studied by Hunault *et al.,* (1995). Explants were maintained for 4 weeks on induction medium containing 2,4-D and kinetin before transferring to a medium devoid of these growth regulators to allow embryo development. The addition of autoclaved or filter-sterilized GA3 to the induction medium or to the embryo development medium increased the number of embryogenic explants. No positive effect was observed when GA3 was added to the micropropagation medium of the mother plantlets. GA3 also counteracted the inhibiting effect of continuous light on the number of embryogenic explants. Moreover, the embryogenic frequency of petiole explants from several genotypes previously considered as poorly reacting was highly enhanced by GA3.

The use of cytological and, especially, nuclear DNA and cpDNA molecular analyses in assessing the genetic stability or instability of organogenesis and somatic embryogenesis-derived plants of one fennel (*Foeniculum vulgare* Mill.) donor plant belonging to Francia Pernod population, was evaluated by Bennici, *et al.,* (2004). Different morphogenic callus types and both organogenic and embryogenic plants showed the normal diploid chromosome number. All regenerated plant types, when examined by RAPD analyses did not show any nuclear DNA polymorphism. No variation was detected in these plants independently from their morphogenic origin in two cpDNA regions which exhibited the same length, base sequence and restriction profiles. Moreover, the study of a cpDNA microsatellite region including a single A/T repeat did not reveal any variation of the repeat numbers in these regenerated plants. These results on the genetic stability and uniformity of organogenic and embryogenic regenerated fennel plants was supported also by a comparison of the cpDNA microsatellite region with other natural fennel plants where variations were found in some of them except Francia Pernod.

Micropropagation is true to type multiplication of selected genotypes using in vitro techniques. MsS medium supplemented with Bap (1.0 mgl^{-1}) and IBa (0.5 mgl^{-1}) was suitable for both multiple shoot formation and callus regeneration. Multiple shoot ranging from 4-7 were produced in 6 weeks in 70% of the cultures. Callusing was observered and callus on transfer to fresh medium regenerated shoots. Plant regeneration was achieved in 60% of the cultures.

Synthetic seeds

Nirmal Babu *et al.,* (1996) and Sajan *et al.*, (1997) reported development of synthetic seeds in Fennel by encapsulating the somatic embryos and sodium alginate. These seeds would be stored successfully with 70% viability under laboratory conditions.

The studies made in the biotechnology of fennel thus can be used for large-scale multiplication of elite genotypes, for production of biomass in the culture vessels, and for creating variability through callus regeneration system by exploring the somaclonal variation to broaden the crop genetic base

Salient Achievements

RF-101: It was developed at S.K.N. College of Agriculture, Jobner (Rajasthan Agriculture University), through recurrent half-sib selection from a local collection of Sohela village (Tonk) and identified for release in 1995. The plants are tall, erect with stout stem. It bears large umbels with long bold grains. It matures in 150-160 days and gives an average yield of 15.5 q/ha.

RF-125: It was developed at S.K.N. College of Agriculture, Jobner (Rajasthan Agriculture University), through recurrent half-sib selection from an exotic collection EC-243380. Introduced from Italy and identified for release in 1997. The plants are early, short structure with compact umbels. It matures in 110-130 days and gives 17.3 q of seed yield per hectare.

RF-143: It was developed at S.K.N. College of Agriculture, Jobner (Rajasthan Agriculture University), through recurrent selection based on individual plant progeny and recommended for the Rajasthan state. Very suitable for the loamy black cotton soils. Medium duration type, medium tall. Having an average yield of 1200 Kg/ha and oil content 1.875%.

RF-178: It was developed through recurrent selection based on individual plant progeny (half sib) from F2 generation of a c ross between UF-125 X UF-133. The plants are erect, medium tall, with bold and attractive seeds, having high volatile oil. It matures in 130-140 days, average yield 16 q/ha.

RF-205: It was developed through recurrent selection based on individual plant progeny (half sib) from F2 generation of a c ross between JF-25 X RF-125. The plants are erect, medium tall, with bold and attractive seeds, having 2.48 % volatile oil. It matures in 140-150 days, average yield 16 q/ha.

RF-145: It was developed through recurrent selection based on individual plant progeny (half sib) from a Local collection of Sidhpur (Sirohi).The plants are erect, medium tall, with bold and attractive seeds, having high volatile oil. It matures in 130-140 days, average yield 21.87 q/ha.

RF-281: It was developed through recurrent selection based on individual plant progeny (half sib) from F2 generation of a c ross between UF(M)-1 X JF-25.The plants are erect, medium tall, with bold and attractive seeds, having high volatile oil. It matures in 130-140 days, average yield 18.25 q/ha.

Gujarat Fennel-1: It was developed by Gujarat Agricultural University through selection from Vijapur Local and released in 1985. The plants are tall and bushy with oblong , medium bold and dark green seeds. It matures in 225 days and gives an average yield 16.95q/ha.

Gujarat Fennel-2: It was developed by Gujarat Agricultural University through selection in local material and therefore yield superiority is not more than 20% over the local varieties. It is tall and long duration variety. Average yield is 18 q/ha.

CO-1: It was developed by TNAU through reselection from PF-35 and released in 1985. The plants are medium structure. It matures in 220 days and average yield 5.67 q/ha in Tamil Nadu.

Azad Saunf-1: Medium plants, resistant to blight and root rot diseases. Escapes attack of aphids due to early maturity. Seeds are bold green with seed yield of 1500 kg/ha.

Pant Madhurika: Tall robust erect plant big umbels having bold seeds with green fine ridges sweet in taste, medium maturity, with seed yield of 1650 kg/ ha.

8.13 Fenugreek Breeding

Breeding Objectives

- To produce high yielding varieties.
- To improve the nutraceutical poperty of fenugreek.
- To produce disease resistance crops.
- To achieve high oleoresin content.
- To get wider adaptability in different environments.

Mutation breeding

A mutation breeding study, using Tristar as base population, was initiated in the green house at LRC to look for mutants with desirable and beneficial phenological traits like determinate growth habit and/or high seed yield. EMS was used as the mutagenic agent at the level of 10, 20,30, 40, 50, 100, 150, 200 and 300 mM concentrations. The seed were pre-soaked in water for 2, 4,6, 8, 12, 16 and 24 h before applying different concentrations of EMS. Treated seed were planted individually in pots containing soil-free mix and were designated as M plants. After 85 d in the greenhouse set to cycle 16 h long days (22°C) and8 h night (15°C) the plants were desiccated with 0.4% Reglone solution. The plants were then allowed to dry for 10 d before separating the seed for yield determination. Seeds from selected M1 plants were again seeded in pots and allowed to produce M seed.

Table 8.12: Salient Breeding achievements of Fenugreek

Sl No.	Variety	Pedigree/ Parentage	*Av. yield (kg/ha)	Duration (days)	Salient features
1	Co.1	Reselection from TG-2356 introduced for North India	680	80-85	A quick growing, dual purpose, early maturing variety tolerant to root rot disease. Seeds contain 21.7% protein.
2	Co 2	Selection from CF 390	480	85-90	Short duration dual purpose variety, field tolerant to Rhizoctonia root rot disease, suitable for both kharif and rabi season. Early maturity, short duration.
3	Rajendra Kanti	Pure line selection from Reghunathpur collection	1300	-	Medium sized bushy plant; early maturity, suitable for intercropping in kharif and rabi season, field tolerant to cercospora leaf spot, powdery mildew and aphids.
4	RMt.1	Pure line selection from Nagpur local	1400	140-150	Vigorous semi erect medium sized, moderately branched growth habit, medium sized, bold and attractive typically yellow coloured grains, moderately resistant to root knot nematode and powdery mildew and aphids

Contd.

Sl No.	Variety	Pedigree/ Parentage	*Av. yield (kg/ha)	Duration (days)	Salient features
5	Lam Sel.1	Selection from germplasm collection of Uttar Pradesh	740	-	Dual purpose varieties, early maturing, bushy type and medium height940cm0, more number of branches and green matter. When cultivated for green leaf purpose it gives an average green yield of 12 tons per hectare. Field tolerant to major pests and diseases.
6	Hisar Sonali	Pure line selection from germpalsm	1700	140-145	Tall and bushy vigorous growing variety, dual purpose variety, late maturity (140-145 days), suitable for cultivation under irrigated condition. Moderately resistant to root rot and aphids.
7	Hisar Suvarna	Pure line selection from local germplasm	1600	130-140	A quick growing, erect and tall, dual purpose, medium maturity (130-140days), moderately resistant to cercospora and powdery mildew. Wider adaptability, suitable for cultivation throughout the country.
8	Hisar Madhavi	Pure line selection from local germplasm of UP	1900	130-140	A quick growing, erect and tall, dual purpose, medium maturity (130-140days), moderately resistant to powdery mildew and to downy mildew. A variety with under adaptability suitable for both irrigated and rainfed condition.
9	Hisar Muktha	Pure line selection natural green seed coated mutant line from UP	2000	135-140	A quick growing seed type variety, medium maturity (135-140days), moderately resistant to powdery mildew and to downy mildew. Erect and tall plants. Wide adaptability. Suitable for both irrigated and rainfed condition

Contd.

Sl No.	Variety	Pedigree/ Parentage	*Av. yield (kg/ha)	Duration (days)	Salient features
10	RMt 303	Mutation breeding from variety RMt 1	1900	140-150	Medium maturity variety (145-150 days) seeds bold ,with typical yellow colour ,less susceptible to powdery mildew
11	RMt305	Mutation breeding from variety RMt 1	1800	120-130	First determinant type, multipodant, early maturing, wider adaptability, resistant to powdery mildew and root knot nematodes. Seeds bold, attractive and yellow, duration 120-125 days.
12	GujMethi 1	Recurrent selection based on pure line selection from J. Fenu 102	1864	-	The first variety from Gujarat released for the state. Plant dwarf.
13	RMt143	Pure line selection	1600	140-150	Moderately resistant to powdery mildew, seeds bold yellow colour, suitable for heavier soils.
15	Pant Ragini	Selection from local germplasm	1200	170-175	A dual purpose tall bushy type resistant to downy mildew and root rots, medium maturity. Seed contain 2-2.5% essential oil
16	NRCS-AM-1 (AM-01-35)	Selection from local germplasm	1720	-	Dual purpose, tolerant to powdery mildew
17	Ajmer Fenugreek-5	ICAR-NRC on Seed spices, Ajmer	-	-	High quality, high yield (17.21 q/ha), high antioxidant (66.43 mg/ BHTE/ppm), suitable for green leaf production in summer undedr protected condition .
18	RMt-351	Mutation breeding from variety RMt 1	1841	140-150	The plants are medium dwarf, bold and attractive seeds, more number of branches and seeds per pod

Contd.

Sl No.	Variety	Pedigree/ Parentage	*Av. yield (kg/ha)	Duration (days)	Salient features
19	RMt-361	Mutation breeding from variety RMt 1	1770	130-140	The plants are erect and bushy, bold and attractive seeds, more number of branches and seeds per pod
20	RMt-361	Mutation breeding from variety RMt 1	2108	130-140	The plants are erect and bushy, bold and attractive yellow-coloured seeds, more number of branches and seeds per pod

8.14 Chilli Breeding

Breeding objectives

a. To get higher yield of fresh vegetables.

b. For higher pungency.

c. For higher oleoresin content.

d. To early fruiting.

e. For development of superior quality.

i) Superior colour and shape of the fruit.

ii) Better development of capsicine, carotine, oleoresin and ascorbic acid.

f. To develop resistance/tolerance to various biotic and abiotic stress.

g. To develop cultivars having wider adaptability.

h. Exploitation of hybrid vigour.

Breeding methods

Introduction and selection

Through introduction and selection, many varieties of chillies have been developped in India particularly 'G' series of Lam Research station (AP), 'L' series of Pant Nagar (Uttarakhand), 'CO' series of Coimbatore and 'K' series of Kavilpatti (Tamil Nadu). In these types, the heavy yielding and high capsaicin character has been united (Muthukrishnan *et al*; 1993). The well-known varities developed through these methods are: G1, G_2, G_3, G_4 NP46A, K1, CO1, Musalwadi, Sindhur and Patna Red (Ram, 1998).

Single plant selection

Typical sub pollinated crops like chilli are prone to some degree of natural cross pollination (16.5%) reported by Muthukrishnan *et al.*, (1993) and as a consequence recombination to genes in the population is possible. Natural out-crossing, spontaneous ,mutation and mechanical mixture supply source material for single plant selection.

Pure line selection

Pure line theory furnishes a basis for the isolation of types that differ appreciably in heritable characters. Progeny of self-pollinated crops breed relatively true, hence the pure line may be homozygous with respect to the observed characters. Pure line selection method requires detailed pedigrees record of each generation.

Forty-eight chilli hybrids, developed by crossing 4 genic male sterile lines and 12 male parents in a line x tester mating design, were grown and evaluated by Patel *et al.*, (2004) for hybrid vigour, gene action and combining ability for 8 traits (days to flower, plant height, primary branches per plant, fruits per plant, fruit length, fruit weight, fruit girth and green fruit yield per plant). Out of 48 hybrids, 7 were good specific combiners for green fruit yield and important yield-contributing traits. Of these, ACMS-2 x RHRC-Pendent, ACMS-4 x ACS 2000-3, ACMS-4 x ACG-77 and ACMS 4 x Punjab Guchhedar also showed significant positive estimates for heterobeltiosis for green fruit yield. ACMS-4 x Punjab Guchhedar showed maximum heterosis for most traits.

The genetic purity of 3 F1 chilli hybrids (C. annuum 'Bangchang' x C. chinense 'CM021', C. chinense 'CM021' x C. chinense 'CM022' and C. annuum '83-168' x C. annuum 'KKU-cluster') was determined using 2 molecular marker techniques, RAPD and ISSR. RAPD analysis successfully detected all 3 F1 hybridity, while ISSR detected only 2. This was due to the RAPD marker system producing a greater number of markers than the ISSR system (Mongkolporn *et al.*, 2004).

An experiment was conducted on chilli cultivars Bhagyalakshmi (G-4), Kovilpatti (K-2), Madurai-1 (MDU-1) and Coimbatore-2 (CO-2) to develop compact, clustered, upright and destalked chilli fruit types to minimize the labour cost involved in harvesting and processing, and to increase the yield due to incorporation of cluster fruit characters. The hybrids obtained from the cross between G-4 and KAU-cluster chilli produced upright, clustered, lengthy fruits with appealing colour. The derivatives obtained from K-2 x KAU-cluster were compact plant types, pendulous clusters with easy detachment, while MDU-1 x KAU-cluster expressed only compact plant type and easy detachment. The results of a related study on the quantitative characters are also discussed (Dhamayanthi and Reddy, 2003).

Heterosis breeding

Appreciable heterosis with respect to yield and various yield attributing characters vas been reported.Twelve F1 hybrids consisting of direct and reciprocal crosses involving four parents (P1: CA 100; P2: Jayant; P3: RHRC and P4: CA-133) were raised by Kanthaswamy *et al.*, (2003). P4 recorded the highest yield (73 g) followed by P1 (68.00 g). Hybrid P4 x P1 was also superior, registering 99.25 g. The direct cross P1 x P4 recorded 49.48 g. P4 x P1 had the highest heterosis in heterobeltiosis, and the per se performance also proved to be high for capsaicin content and dry fruit yield per plant. P4 x P1 exhibited the highest heterosis percent, per se performance and photosynthetic rate. Generally, high yielding genotypes had high net photosynthetic rate. Dry matter production was largely contributed by photosynthetic rate (12.28 g plant-1) in P4 x P1. Dry matter production exhibited negative heterosis in all hybrids. Crop growth rate and net assimilation rate were positively correlated with yield. P2 x P4 exhibited the highest photosynthetic rate by the heterosis percent and per se performance.

A set of 21 male sterile based F1 chilli hybrids developed as a result of line x tester mating design involving 3 cytoplasmic genic male sterile (CGMS) lines and 7 testers were evaluated by Shankarnag *et al.*, (2006) for growth (stem girth, plant height, plant spread, number of secondary and tertiary branches), earliness (days to first and 50 per cent flowering) and early green fruit yield. The cross L5 x T14 was most heterotic over both the control hybrids for early green fruit yield followed by L3 x T14.

Shankarnag *et al.*, (2005) earlier worked on exploitation of heterosis using male sterility in chilli for fruit and yield parameters. Three cytoplasmic genic male sterile lines were crossed with seven testers to develop 21 F1 hybrids. The promising crosses for the trait total green fruit yield with high significant economic heterosis in the order of their merit were L5 x T14 (56.51%), L3 x T14 (47.76%) and L5 x T15 (47.43%).

A set of 36 F1 hybrids produced from a 9x9 half diallel mating design involving one bell pepper cultivar (Arka Gaurav) and eight hot pepper advanced breeding lines (VR-1, VR-2, VR-42, VR-14, VR-17, VR-27, VR-47 and VR-55) were evaluated by Prasad *et al.*,(2003). Data on the magnitude of heterosis revealed that VR-2 x VR-55 was superior over best parent and standard control with respect to dry fruit yield per plant. VR-1 x VR-2 was superior over best parent and standard control with respect to number of fruits per plant.

Heterosis and combining ability effects for growth parameters (stem girth, plant height, height at first branching, plant spread, number of primary branches, and number of secondary branches) were studied by Linganagouda *et al.*, (2003) in 27 F1 hybrids and 12 parental lines (3 Capsicum lines as female parents and 9 chilli lines as male parents). Heterosis for plant height and number of secondary branches was attributed to the additive and non-additive components of genetic variance. For plant spread and number of primary branches, heterosis was attributed to the non-additive and additive components of genetic variance, respectively.

Heterosis, hetrobeltiosis and standard heterosis were studied by Gondane and Deshmukh (2004) in thirty-three chilli hybrids obtained by crossing three male sterile lines viz., CA-960, Jwala, AKC-86-25 with eleven pollen parent (testers) viz., GP-89, GP-90, GP-93, GP-219, GP-234, GP-312, GP-313, GP-314, GP-315 and GP-361. The cross combinations Jwala x GP-90, AKC-86-25 x GP-313, CA-960 x GP-22 and AKC-86-25 x GP-314 were the best crosses on the basis of high heterosis, heterobeltiosis and standard heterosis signifying commercial exploitation of heterosis from these creosses.

Breeding for disease and insect-pest resistance

Anthracnose, bacterial leaf spot and virus complex (TMV, CMV, PVX and PVY) are important diseases of chillies in India. Patil *et al.*, (2003) identified the resistant donor chilli (*Capsicum annuum*) genotype against fruit rot caused by *Colletotrichum capsici*, which may be useful in the management of the disease. One and half-month-old seedlings of 20 cultivars of chilli, raised on seedbed treated with phorate granules 10% at 10 kg/ha, were used for transplanting in a single row of 6-m length. The observations on the incidence of fruit rot were recorded at each picking. None of the cultivars was found to be immune to the fruit rot under Vidarbha conditions. However, cultivars Jayanti and Phule Jyoti appeared to be resistant and 11 cultivars, namely Jwala, Phule Suryamukhi, Arka Lohit, KAC-86-25, Agni Rekha, AKC-BC-89-11, Parbhani Tall, X-235, AKC-BC-89-8, G-4 and Surkta, were found to be moderately resistant. The remaining 7 genotypes were susceptible to the disease.

Genetic nature of cucumber mosaic virus (CMV) resistance was studied in three hot pepper crosses, C1024 (resistant) x CA87067 (susceptible), C1024 x Chilli (susceptible) and C1034 (resistant) x CA87067. CMV resistance was controlled by a nuclear recessive simple gene. Broad and narrow sense heritability estimates were high Herison *et al.*, (2004). Three hundred and seven genotypes belonging to four cultivated and one wild species of Capsicum were screened against pepper leaf curl virus (PepLCV) causing devastating leaf curl disease of chilli (*Capsicum annuum*) by Kumar *et al.*, (2006). Initial screening was done under field conditions based on coefficient of infection (CI), disease reaction to each genotype was assigned. Subsequently, selfed progenies of eight symptom-less and highly resistant lines were challenged by viruliferous white fly under glasshouse conditions, out of which only three genotypes, viz. GKC-29, BS-35 and EC-497636 showed no symptom. Using scion and root stalk of susceptible genotype (Pusa Jwala), these three putative symptoms-less genotypes were further challenged by grafting and alternate grafting. The resistant reactions of GKC-29, BS-35, EC-497636 were confirmed because even after 50 days of successful grafting/alternate grafting, no viral symptom appeared on all the grafted plants of these genotypes. When subjected to PCR amplification with degenerate primers deigned to detect gemnivirus like PepLCV, the three symptom-less genotypes did not show any amplification, suggesting that the resistant reaction in three identified symptom-less resistant sources was because of the absence of viral genome and they are not symptom-less carrier.

Laboratory and greenhouse studies were conducted by Kaur *et al.*, (2005) to determine the effective sources of resistance against different races of Colletotrichum capsici in 71 chilli germplasm lines. Freshly harvested ripened fruits from healthy plants were pin-pricked prior to inoculation and then inoculated by placing uniform drop of spore suspension. Seventy-one germplasm lines evaluated against 37 *C. capsici* isolates showed that most of the germplasm lines were susceptible to almost all the isolates and only few exhibited resistant or moderately resistant reaction to certain isolates. Cultivars California Wonder, Surajmukhi and S-2530 showed resistance to nine isolates with a severity index of 2.1-5.0%. Thirty-three local lines from Himachal Pradesh were mostly susceptible to different isolates.

reported that the incidence of chilli leaf curl virus (CLCV) in 13 chilli (*Capsicum annuum*) cultivars (Suryamukhi, NP-46A, G-4, IR-8, Akashi, Dinhata Local-1, Dinhata Local-2, Beldanga Local, Gosaba Local, Bullet, Dinajpur Local, Pusa Jwala and Pusa Sadabahar) was studied by De *et al.*, (2005) in Kalyani, West Bengal, India. Thirty-day-old seedlings were transplanted in the last week of February. Plants remained disease-free up to 15 days after transplanting (DAT). In most of the cultivars, disease incidence was initially recorded at 45 DAT. At 60 DAT, CLCV was observed in all cultivars, and the incidence increased until 105 DAT. In most of the cultivars, CLCV incidence increased at an increasing rate up to 75 DAT, then increased at a decreasing rate or stabilized thereafter. The initial infection was most severe in Dinajpur local (6.67 and 7.14 in 2000 and 2001, respectively), and least severe in Pusa Jawala in 2000 (1.87%) and IR-8 in 2001 (2.57%). The number of days to first disease appearance was highest in IR-8 (55-56 days) and lowest in Dinajpur local (22-24 days). In other cultivars, disease symptoms were evident in 26-5 days. The number of days required to reach maximum infection was lowest in Dinajpur local (34-37 days), and highest in Bullet in 2000 (62 days) and Akashi in 2001 (59 days). The rate of spread of CLCV disease per day was highest in Dinajpur local (0.784 and 0.0752) and lowest in IR-8 (0.0415 and 0.416). Dinajpur Local was highly susceptible to CLCV, whereas Akashi and Pusa Jwala were less susceptible.

Mutation breeding

Mutants like X-rays, Gamma-rays, EMB, MNH have been used in chillies induction of mutations (Desmukh *et al.*, 1982; Ram, 1998). But so far very few varieties, eg., MDU-1 has been released for commercial cultivation. This variety was developed by treating the seeds of verities K-1 by gamma- rays. It has compact plant type, higher yield and capsaicin content.

Table 8.13: Salient Breeding achievements of Chilli

Name of Variety/ Hybrid	Neme of Institute	Year of release	Important features
Punjab Lal	PAU, Ludhiana, Punjab	1985	Dwarf plant, bushy, erect fruit, rich in Capsaicin (0.7 %), oleoresin, dry matter, resistant to TMV
Pusa Jwala	IARI, New Delhi	1974	Dwarf plant, bushy, fresh fruit yield 75 q/ha,resistant to thrips, mite and aphids
Bhagya Lakhsmi (G4)	Agric. Res. Sta., Lam, Guntur	1968, 1977	Narrow leaves, avg. dry fruit yield 11 q/ha (rainfed) and 50 q/ ha (irrigated)
Andhra Jyoti (G5)	Agric. Res. Sta., Lam, Guntur	1972, 1977	Fruits short, stout and conical shaped
Pant C-1	GBPUA and T Pantnagar, Uttarakhand	-	Short stature, fruits borne upright in cluster, highly pungent, tolerant to mosaic and leaf curl
Pusa Sada Bahar	IARI, New Delhi	1989	Plants erect, fruits erect, in cluster, resistant to CMV, TMV and leaf curl
NP46 A	IARI, New Delhi	-	Dwarf plant, bushy, fresh yield 70-80 q/ha,tolerant to thrips, susceptible to virus
MDU 1	Agric. College and Res. Instt, Madurai, TN	1975	Plants dwarf, compact, suitable for high density planting, high capsaicin, dry yield 19 q/ha
Surjamukhi	HPKV, Paklampur	-	Plants erect, branches arises from base, fruits erect, highly pungent, fresh fruit yield 80-100 q/ha, dry fruit yield 12-15 q/ha

Colour Plates

Photographs of Spice Breeding Achievements

A:IISR Girimunda **B**: Sreekara **C**: Subhakara **D**: Pournami **E**: Panchami **F**:IISR Thevam

Photographs of Spice Breeding Achievements

Malabar (Prostrate panicle)

Mysore (Erect panicle

Vazhukka (Semi-erect Panicle)

A: IISR Malabar Excel(Bl Pepper) **B:** Malabar(Small Cardamom) **C:** Mysore (Small cardamom), **D:** Vazhukka (Small Cardamom)

Photographs of Spice Breeding Achievements

A: CCS I(short plant type) **B:** Green Gold **C:** IISR Avinash(rhizome rot resistant) **D:** IISR Vijetha (katte resistant) **E:** Protoplast culture **F:** Variability in Mace colour

Photographs of Spice Breeding Achievements

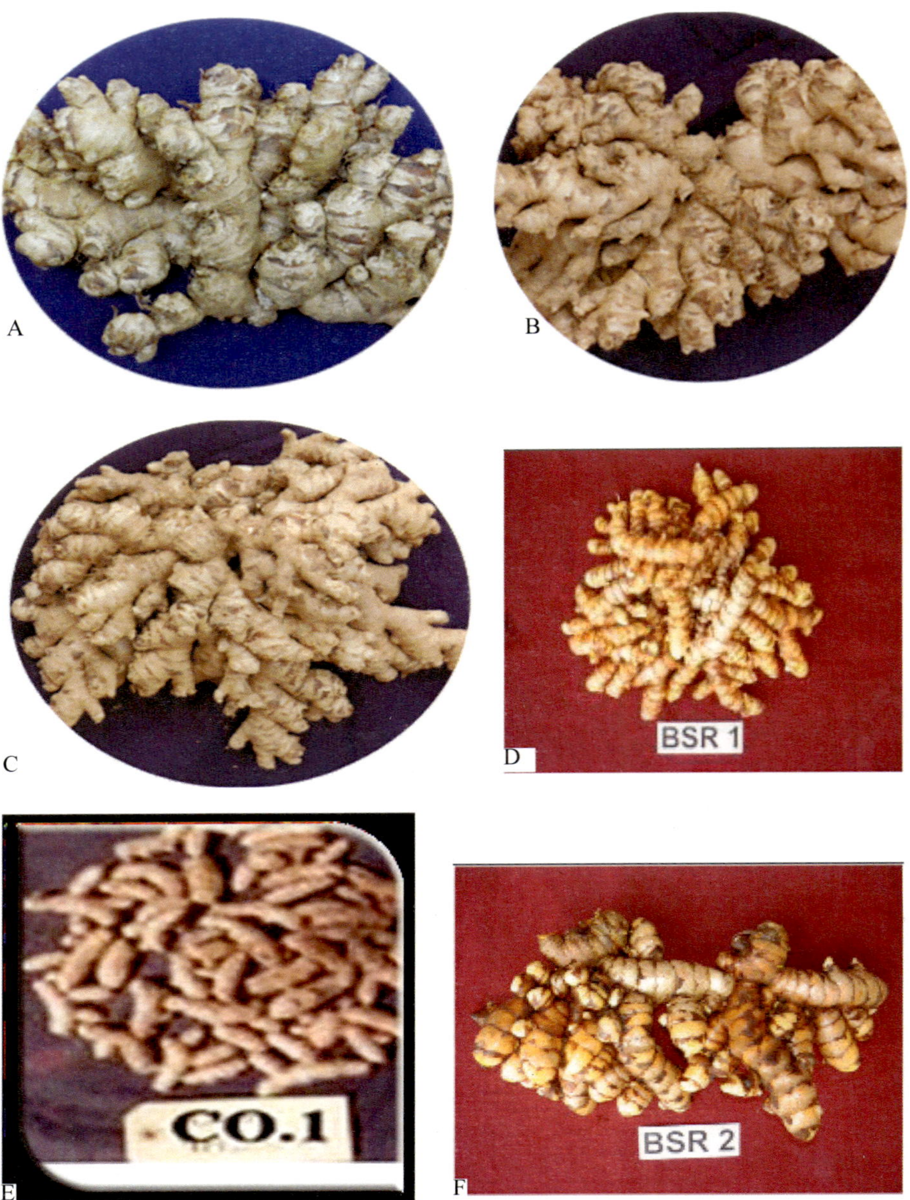

A: IISR- Mahima Ginger **B:** IISR- Rejatha Ginge **C:** IISR- Varada Ginger **D:** BSR-1 Turmeric **E:** Co-1 Turmeric **F:** BSR-2 Turmeric

Photographs of Spice Breeding Achievements

A and B: Induction of micro rhizomes and rhizomes developed from them **C:** Root knot nematodes infection in turmeric **D:** Lesion nematodes infection in turmeric

Photographs of Spice Breeding Achievements

A B C D

A: Nutmeg:Konkan Sugandha **B:** Nutmeg: Konkan Swad **C:**Nutmeg:Konkan Shrimanti **D:** Nutmeg:IISR Vishwashri

Photographs of Spice Breeding Achievements

Fig 1: *In vitro* regeneration of garlic plants.

A: Protocorms regenerated from root tips, **B-C:** Protocorms, **D:** Microbulbil, **E -F:** Whole plant.

Fig 2: Development of transgenics in Black pepper

Chapter 2: Morphology of Spice Crops

Fig. 2.6[A-F]: A, An individual flower of Amomum subulatum. **B,** Anther–stigma column **C,** Same as B, but after repeated visits by the honey bee. There is hardly any pollen left on the anther. **D,** The bumble-bee, *Bombus haemorrhoidalis* **E,** The honey bee, Apis cerana collecting pollen from the flower. **F,** A flower to show significant increase in the distance between the anther–stigma column and labellum.

M. beddomi *M. malabarica*

M. fragrans

Fig. 2.7: Different species of Myristica

Fig. 2.8: Flower and different parts

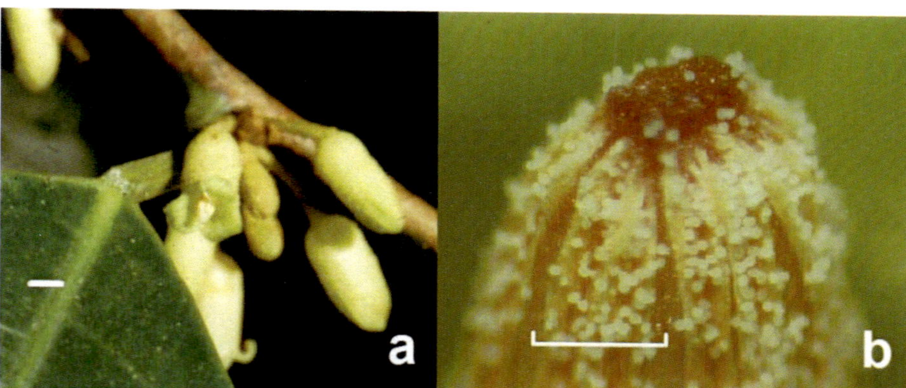

Fig. 2.10: Male flowers and fully opened anther

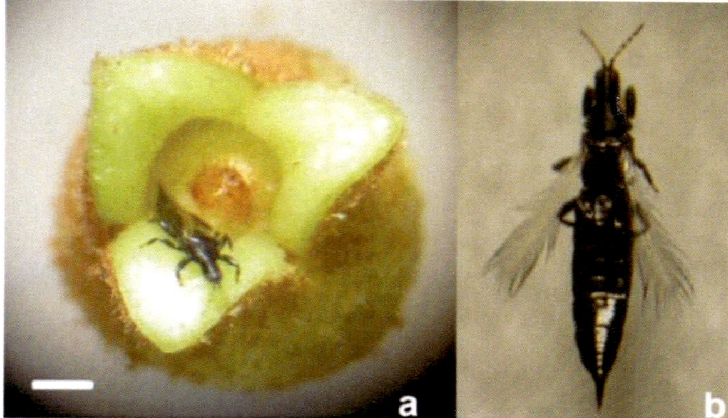

Fig. 2.11: Floral vectors

Chapter 8: Global Breeding Research on Important Spice Crops

Fig. 8.1: Nutmeg tree with mature fruits

References

Aarthi, S., J. Suresh and D. Prasath. 2018. Morphological characterization of Indian turmeric (*Curcuma longa* L.) genotypes using DUS descriptor. Journal of Plantation Crops. 46: 173-179.

Abbas, M. S. T. 2018. Genetically engineered (modified) crops (*Bacillus thuringiensis* crops) and the world controversy on their safety. Egyptian Journal of Biological Pest Control. 28:52.

Abdallah, S. A. S., M.Y.A. Yakoup and M.Y.H. Abdalla. 2017.Micropropagation of oregano (*Origanum syriacum* L.) through tissue culture technique. Journal of Plant Production. 8:635-639.

Abhay D., R . K. Sharma, E. V. D. Sastry and D.Singh. 2000. Combining ability analysis in varietal diallel cross of fennel (*Foeniculum vulgare* Mill.) for yield and yield related traits. Spices and aromatic plants: challenges and opportunities in the new century. Contributory papers Centennial conference on spices and aromatic plants, Calicut, Kerala, India,78-81.

Acharya, B. C. and M.V. Ramji. 1977. Experimental androgenesis in plants—a review. Proceedings of the Indian Academy of Sciences-Section B. 86 (6):337-360).

Acquaah, G. 2015. Conventional plant breeding principles and techniques, p. 115-158.In J. M. Al-Khayri, S. Jain and D. V. Johnson, eds. Advances in Plant Breeding Strategies: Breeding, Biotechnology and Molecular Tools.Springer, Cham, New York.

Aebersold, R. and B.F. Cravatt.2002.Proteomics–advances, applications and the challenges that remain. Trends in Biotechnology.20:PS1-PS2.

Aghaei, Z., M. Talebi and M. Rahimmalek. 2017. Assessment of genetic diversity within and among sage (*Salvia*) species using SRAP markers. Plant Genetic Resources. 15: 279-282.

Aghighiravan, F., M. Shokrpour, V. Nazeri and M.R. Naghavi. 2019. Phylogenetic Assessment of Some Species of Crocus Genus Using DNA Barcoding. Journal of Genetic Resources. 5: 118-129.

Ahmad N., B. Guo, H. Fazal, B. H.Abbasi, L. Chun-Zhao, T.Mahmood and Z. K. Shinwari. 2011. Feasible plant regeneration in black pepper from petiole explants. Journal of Medicinal Plants Research. 5: 4590-4595.

Ahmad, N., B.H. Abbasi, H. Fazal, M.A. Khan and M.S. Afridi. 2014. Effect of reverse photoperiod on *in vitro* regeneration and piperine production in *Piper nigrum* L. Comptes Rendus Biologies. 337: 19-28.

Ahmad, N., B.H. Abbasi, I. Rahman, and H.Fazal. 2013. *Piper nigrum*: micropropagation, antioxidative enzyme activities, and chromatographic fingerprint analysis for quality control.Applied Biochemistry and Biotechnology. 169: 2004-2015.

Ahmad, N., H. Fazal, B.H. Abbasi, Md.Rashid, A.Tariq and F. Nighat. 2010. Efficient regeneration and antioxidant potential in regenerated tissues of *Piper nigrum* L. Plant Cell Tissue and Organ Culture (PCTOC). 102: 129-134.

Ahn, Y. K., S. Karna, T.H. Jun, E.Y. Yang, H.E. Lee, J.H. Kim and J.H. Kim. 2016. Complete genome sequencing and analysis of *Capsicum annuum* varieties. Molecular Breeding. 36:140.

Aizat, W. M., D.A. Dias, J.C. Stangoulis, J.A. Able, U. Roessner and A.J. Able. 2014. Metabolomics of capsicum ripening reveals modification of the ethylene related-pathway and carbon metabolism. Postharvest Biology and Technology.89: 19-31.

Aizat, W. M., J. A. Able, J. C.Stangoulis and A. J. Able. 2013. Proteomic analysis during capsicum ripening reveals differential expression of ACC oxidase isoform 4 and other candidates. Functional Plant Biology. 40:1115-1128.

Ajayi, O. B., S.F. Akomolafe and F.T. Akinyemi. 2013. Food value of two varieties of ginger (*Zingiber officinale*) commonly consumed in Nigeria. ISRN Nutrition. 2013.

Ajith A., M. Parani, C.S. Rao, R. Latha and P. Balakrishna. 1997. Micropropagation and genetic fidelity studies in *Piper longum* L. In: Biotechnology of Spices, Medicinal and Aromatic Plants. Edison S., K.V. Ramana, B. Sasikumar, K.N. Babu, S.J. Eapen (eds.), Indian Society of Spices, Calicut, Kerala, India.

Akula, R. and G.A.Ravishankar. 2011. Influence of abiotic stress signals on secondary metabolites in plants. Plant Signaling & Behavior.6:1720-1731.

Ali, A. M. A., M. E. M. El-Nour and S. M. Yagi. 2018b. Total phenolic and flavonoid contents and antioxidant activity of ginger (*Zingiber officinale* Rosc.) rhizome, callus and callus treated with some elicitors. Journal of Genetic Engineering and Biotechnology. 16: 677-682.

Ali, A., M. Khan, R. Sharif, M. Mujtaba and S.J. Gao. 2019. Sugarcane Omics: An update on the current status of research and crop improvement. Plants. 8: 344.

Ali, M., A. Mujib, N. Zafar and D. Tonk.2018a. Somatic embryogenesis, biochemical alterations and synthetic seed development in two varieties of coriander (*Coriandrum sativum* L.). Advances in Horticultural Science.32: 239-248.

Ali, M., A.Mujib, D. Tonk and N. Zafar. 2017. Plant regeneration through somatic embryogenesis and genome size analysis of *Coriandrum sativum* L. Protoplasma. 254: 343-352.

Al-Jibouri, A. M. J., A.S. Abd, D.M. Majeed and E.N. Ismail. 2012.Influence of abiotic elicitors on accumulation of thymol in callus cultures of *Origanum vulgare* L. Journal of Life Sciences. 6:1094-1099.

Altpeter, F., N.M. Springer, L.E. Bartley, A.E. Blechl, T.P. Brutnell, V. Citovsky,.....and P.G. Lemaux. 2016. Advancing crop transformation in the era of genome editing. The Plant Cell.28: 1510-1520.

Alvares Bianchi, P., L.Renata Almeida da Silva, A. André da Silva Alencar, P.Henrique Araújo Diniz Santos, S.Pimenta, C.Pombo Sudré and R. Rodrigues. 2020. Biomorphological Characterization of Brazilian *Capsicum Chinense* Jacq. Germplasm. Agronomy. 10: 447.

Anami, S., E. Njuguna, G. Coussens, S. Aesaert and M. Van Lijsebettens. 2013. Higher plant transformation: principles and molecular tools. International Journal of Developmental Biology. 57:483-494.

Anamika, K., S. Verma, A. Jere and A. Desai. 2016. Transcriptomic Profiling Using Next Generation Sequencing-Advances, Advantages, and Challenges. Next Generation Sequencing-Advances, Applications and Challenges.9: 7355-7365.

Anasori, P. and G. Asghari. 2009. Effects of light and differentiation on gingerol and zingiberene production in callus culture of *Zingiber officinale* Rosc. Research in Pharmaceutical Sciences.3: 59-63.

Anjali V., K. Nikhil and S.A. Ranade. 2004. Genetic diversity amongst landraces of a dioecious vegetatively propagated plant, betelvine (*Piper betle* L.). Journal of Biosciences. 29: 319-328.

Antcheva, N., A, Patthy, A, Athanasiadis, B. Tchorbanov, S. Zakhariev and S. Pongor. 1996. Primary structure and specificity of a serine proteinase inhibitor from paprika (*Capsicum annuum*) seeds. Biochimica et Biophysica Acta (BBA)-Protein Structure and Molecular Enzymology. 1298:95-101.

Anu, A. and K.V. Peter. 2003. Analysis of seed protein of 29 lines of *Capsicum annuum* L. by polyacrylamide gel electrophoresis. Genetic Resources and Crop Evolution.50: 239-243.

Aranha, B. C., J.F. Hoffmann, R.L. Barbieri, C.V. Rombaldi and F.C. Chaves. 2017. Untargeted metabolomic analysis of Capsicum spp. by GC–MS. Phytochemical Analysis. 28: 439-447.

Ari, E., T. Yildirim, N. Mutlu, S. Büyükalaca, Ü. Gökmen and E. Akman. 2016. Comparison of different androgenesis protocols for doubled haploid plant production in ornamental pepper (*Capsicum annuum* L.). Turkish Journal of Biology.40:944-954.

Aryakia, E., H.R. Karimi, M.R. Naghavi and S.A.S Fazeli. 2016. Morphological characterization of intra-and interspecific diversity in some Iranian wild *Allium* species. Euphytica.211: 185-200.

Asamenew, G., H.W. Kim, M.K. Lee, S.H. Lee, Y.J. Kim, Y.S. Cha, and J.B. Kim. 2019. Characterization of phenolic compounds from normal ginger (*Zingiber officinale* Rosc.) and black ginger (*Kaempferia parviflora* Wall.) using UPLC–DAD–QToF–MS. European Food Research and Technology.245:653-665.

Ashokkumara K., M. Murugana, M.K.Dhanyaa and T.D.Warkentin. 2020. Botany, traditional uses, phytochemistry and biological activities of cardamom [*Elettaria cardamomum (L.) Maton*]–A critical review. Journal of Ethnopharmacology .246:112244. doi: 10.1016/j. jep.2019.112244. .

Aslam, B., M. Basit, M.A. Nisar, M. Khurshid and M.H. Rasool. 2017. Proteomics: technologies and their applications. Journal of Chromatographic Science.55:182-196.

Awasthi, P., A. Singh, G. Sheikh, V. Mahajan, A.P. Gupta, S. Gupta, and S.G. Gandhi. 2017. Mining and characterization of EST-SSR markers for *Zingiber officinale* Roscoe with transferability to other species of Zingiberaceae. Physiology and Molecular Biology of Plants. 23(4): 925-931.

Awin, T., A. Mediani, S.W. Leong, S.M.M. Faudzi, K. Shaari and F. Abas. 2019. Phytochemical and bioactivity alterations of *Curcuma* species harvested at different growth stages by NMR-based metabolomics. Journal of Food Composition and Analysis.77:66-76.

Ayed, C., C. Bayoudh, A. Rhimi, N. Mezghani, F. Haouala and B. AL Mohandes Dridi. 2018. *In vitro* propagation of tunisian local garlic (*Allium sativum* L.) from shoot tip culture. Journal of Horticulture and Postharvest Research. 1(2):75-86.

Babaei, S., M. Talebi, M. Bahar and H. Zeinali. 2014. Analysis of genetic diversity among saffron (*Crocus sativus*) accessions from different regions of Iran as revealed by SRAP markers. Scientia Horticulturae. 171: 27-31.

Babu, K. N., G. Yamuna, K. Praveen, D. Minoo, P.N. Ravidran and K.V. Peter. 2012. Cryopreservation of spices genetic resources. p. 457-484, In I. I. Katcov, ed. Current Frontiers in Cryobiology, InTech, Croatia.

Babu, K. N., M.K. Rajesh, K. Samsudeen, D. Minoo, E.J. Suraby, K. Anupama and P. Ritto. 2014. Randomly amplified polymorphic DNA (RAPD) and derived techniques. p. 191-209, In P. Besse, ed. Molecular Plant Taxonomy, Humana Press, Totowa, NJ.

Babu, N., E.J. Suraby, J. Cissin, D. Minoo, T. Pradeepkumar, V.A. Parthasarathy and K.V. Peter. 2013. Status of transgenics in Indian spices. Journal of Tropical Agriculture. 51:1-14.

Babu, R., S. K. Nair, B.M. Prasanna and H.S. Gupta. 2004. Integrating marker-assisted selection in crop breeding–prospects and challenges. Current Science. 87: 607-619.

Bagali, P. G., P.A.D.H. Prabhu, K. Raghavendra, P.G. Bagali, S. Hittalmani and J.S. Vadivelu. 2010. Application of molecular markers in plant tissue culture. Proeeding of Asia Pacific Conference on Plant Tissue and Agribiotechnology (APaCPA) (Vol. 17, p. 21).

Bahadur, V., V. Yeshudas and O.P. Meena. 2016. Nature and magnitude of genetic variability and diversity analysis of Indian turmeric accessions using agro-morphological descriptors. Canadian Journal of Plant Science. 96:371-381.

Balachandran, S. M., S.R. Bhat and K.P.S. Chandel. 1990. *In vitro* clonal multiplication of turmeric (*Curcuma spp.*) and ginger (*Zingiber officinale* Rosc.). Plant Cell Reports.8:521-524.

Banerjee N.S., P. Manoj and M.R. Das. 1999. Male sex-associated RAPD markers in *Piper longum* L. Current Science. 77: 693-695.

Barbosa, G. B., N.S. Jayasinghe, S.H. Natera, E.D. Inutan, N.P. Peteros and U. Roessner. 2017. From common to rare Zingiberaceae plants-A metabolomics study using GC-MS. Phytochemistry. 140:141-150.

Bardakci, F. 2001. Random amplified polymorphic DNA (RAPD) markers. Turkish Journal of Biology. 25(2):185-196.

Baruah, J., S.K. Pandey, T. Begum, N. Sarma, M. Paw and M. Lal. 2019. Molecular diversity assessed amongst high dry rhizome recovery Ginger germplasm *(Zingiber officinale* Roscoe) from NE-India using RAPD and ISSR markers. Industrial Crops and Products.129: 463-471.

Basak, S., V. Kesari, A.M. Ramesh, L. Rangan, A. Parida and S. Mitra. 2017. Assessment of genetic variation among nineteen turmeric cultivars of Northeast India: nuclear DNA content and molecular marker approach. Acta Physiologiae Plantarum. 39 : 45.

Baskaran, P., A, Kumari and J. Van Staden. 2015. Embryogenesis and synthetic seed production in Mondia whitei. Plant Cell, Tissue and Organ Culture (PCTOC). 121:205-214.

Bates, G. W., J.J.Gayno and N.S. Shekhawat. 1983. Fusion of plant protoplasts by electric fields. Plant Physiology.72:1110-1113.

Becker-Andre, M. and K. Hahlbrock. 1989. Absolute mRNA quantification using the polymerase chain reaction (PCR). A novel approach by a PCR aided *t*ranscipt titration assay (PATTY). Nucleic Acids Research.17: 9437-9446.

Bejoy, M., M. Dan, N.P. Anish, A.R. Nair, B.J. Radhika and K. Manesh. 2012. Micropropagation of an Indian ginger (*Curcuma vamana* Sabu and Mangaly): a wild relative of turmeric. Biotechnology.11: 333-338.

Belavadi, V.V. and C.Parvathi.1998. Estimation of honeybee colonies required for effective pollination in cardamom. In : Proc. National symp. Diversity of Social Insects and other arthropods and the functioning ecosystems. III Congress of IUSSI, Mudigere, p. 30.

Belavadi, V.V., Venkateshalu and H.R.Vivek.1977. Significance of style in cardamom corolla tubes for honeybee pollination. Current Science.73: 287- 290.

Belavadi,V.V., C. Parvathi and B.Raju. 2000. Optimal foraging by honeybees on cardamom. Recent Advances in Plantation Crop Research. pp. 318-322.

Beniwal, B.R., S.R. Kumar and B.R. Chaudhary 2007. Phenotypic stability in Isabgol. Annals of Arid Zone. 46: 59-63.

Bennici, A., M.Anzidei, G. G. Vendramin. 2004. Genetic stability and uniformity of *Foeniculum vulgare* Mill. regenerated plants through organogenesis and somatic embryogenesis. Plant Science. 166: 221-227 .

Bernath, J., A. Kattaa, E. Nemeth, R. Franke. 1996. Production-biological investigation of fennel (*Foeniculum vulgare*) populations of different genotype. Atti del convegno internazionale: Coltivazione e miglioramento di piante officinali, Trento, Italy 2 3 giugno, 287-292 .

Beyene, Y., A.M. Botha and A.A. Myburg. 2005. A comparative study of molecular and morphological methods of describing genetic relationships in traditional Ethiopian highland maize. African Journal of Biotechnology. 4: 586-595.

Bhagyawant, S. S. 2016. RAPD-SCAR markers: an interface tool for authentication of traits. Journal of Biosciences and Medicines. 4:1-9.

Bharti, R., S. Kumar and M. J. Parekh. 2018. Development of genomic simple sequence repeat (gSSR) markers in cumin and their application in diversity analyses and cross-transferability. Industrial Crops and Products. 111:158-164.

Bhat A.I., Hareesh, P.S. and R. Madhubala. 2005. Sequencing of coat protein gene of an isolate of Cucumber Mosaic Virus infecting black pepper in India. Journal of Biochemistry and Plant Biotechnology.14:37- 40.

Bhat S.R., Chandel K.P.S. and Malik S.K.1995. Plant regeneration from various explants of cultivated Piper species. Plant Cell Reports.14 : 398 - 402.

Bhatt, J., S. Kumar, S. Patel and R. Solanki. 2017. Sequence-related amplified polymorphism (SRAP) markers based genetic diversity analysis of cumin genotypes. Annals of Agrarian Science. 15: 434-438.

Bhattacharya, M. and A. Sen. 2006.Rapid *in vitro* multiplication of disease-free *Zingiber officinale* Rosc. Indian Journal of Plant Physiology.11: 379-384.

Bhering, L. L. and S. Xu. 2019. Indirect response in genome wide selection using selected markers. Bioscience Joural. 35: 1588-1598.

Bhojwani, S. S. and M.K. Razdan. 1996. Plant tissue culture, theory and practices. A revised edition. Studies in Plant Science. 5. 407-450.

Bianchi, P.A., L.R. A. da Silva, A.A.S. Alencar, P.H.A.D. Santos,.... R. Rodrigues. 2020. Biomorphological characterization of Brazilian *C. chinense* Jacq. Germplasm. Agronomy.10: 447.https://doi.org/10.3390/agronomy10030447

Bischof, S., J. Grossmann and W. Gruissem. 2012. Proteomics and its application in plant biotechnology. p. 55-65, In A. Altman and P. M. Hasegawa, eds. Plant Biotechnology and Agriculture : Prospects of 21st Century. Elsevier.Academic Press.

Blanco, E. Z., M.M. Bajay, M.V.B.M. Siqueira, M.I. Zucchi and J.B. Pinheiro, J. B. 2016. Genetic diversity and structure of Brazilian ginger germplasm (*Zingiber officinale*) revealed by AFLP markers. Genetica.144: 627-638.

Blears, M. J., De Grandis, S. A., H. Lee and J.T. Trevors. 1998. Amplified fragment length polymorphism (AFLP): a review of the procedure and its applications. Journal of Industrial Microbiology and Biotechnology. 21: 99-114.

Bora, G., H.K. Gogoi and P.J. Handique. 2018. Callus mediated *in vitro* regeneration of Naga chilli (*Capsicum chinense* Jacq.): the fiery pepper from North East India. International Journal of Current Microbiology and Applied Sciences.7: 1312-1324.

Bora, G., H.K. Gogoi and P.J. Handique. 2019. Influence of silver nitrate and glutamine on *in vitro* organogenesis of Lota Bhot (*Capsicum chinense* Jacq.), an indigenous pungent pepper variety of Assam. Journal of Applied Biology & Biotechnology. 7:21-28.

Botek, P., J. PouStka and J. Hajslova .2007.Determination of banned dyes in spices by liquid chromatography-mass spectrometry. Czech Journal of Food Science. 25:17-24.

Bourgaud, F., A. Gravot, S. Milesi and E. Gontier. 2001. Production of plant secondary metabolites: a historical perspective. Plant Science.161: 839-851.

Brenner, S., M. Johnson, J. Bridgham, G. Golda, D. H. Lloyd, D. Johnson, and R. Roth. 2000. Gene expression analysis by massively parallel signature sequencing (MPSS) on microbead arrays. Nature Biotechnology.18:630-634.

Brewer, G. J. and C.F. Sing. 1970. Introduction to isozyme techniques. New York-London: Academic Press.

Broome, O.C. and R.N. Zimmerman. 1978. *In-vitro* propagation of black pepper. *Horticultural Science*. 43: 151-153.

Bulle, M., R. Yarra and S. Abbagani. 2016. Enhanced salinity stress tolerance in transgenic chilli pepper (*Capsicum annuum* L.) plants overexpressing the wheat antiporter (TaNHX2) gene. Molecular Breeding. 36: 36.https://doi.org/10.1007/s11032-016-0451-5

Bunnik, E. M. and K.G. Le Roch. 2013. An introduction to functional genomics and systems biology. Advances in Wound Care.2:490-498.

Cai, W. Q., R.X. Fang, H.S. Shang, X. Wang, F.L. Zhang, Y.R. Li,... and K.Q. Mang. 2003. Development of CMV-and TMV-resistant chili pepper: field performance and biosafety assessment. Molecular Breeding. 11:25-35.

Cairns, J. E., Sonder, K., Zaidi, P. H., Verhulst, N., Mahuku, G., Babu, R. and Rashid, Z. (2012). Maize production in a changing climate: impacts, adaptation, and mitigation strategies. p. 1-58, In D. L. Sparks, ed. Advances in Agronomy, Academic Press.

Cardoso, R., C.F. Ruas, R.M. Giacomin, P.M. Ruas, E.A. Ruas, R.L. Barbieri, and L.S. Goncalves. 2018. Genetic variability in Brazilian *Capsicum baccatum* germplasm collection assessed by morphological fruit traits and AFLP markers. PloS One.13(5).https://doi.org/10.1371/journal.pone.0196468

Carlson, P. S., H.H. Smith and R.D. Dearing. 1972. Parasexual interspecific plant hybridization. Proceedings of the National Academy of Sciences.69:2292-2294.

Chahota, R. K., V. Sharma, M. Ghani, T.R. Sharma, J.C. Rana and S.K. Sharma. 2017. Genetic and phytochemical diversity analysis in *Bunium persicum* populations of north-western Himalaya. Physiology and Molecular Biology of Plants. 23(2):429-441.

Chakravarti ,A.K. 1948. Multiplication of chromosome numbers in relation to speciation in Zingiberaceae. Science and Culture, 14:137-140.

Chan, S. N., N. Abu Bakar, M. Mahmood, C.L. Ho and N.A. Shaharuddin. 2014. Molecular cloning and characterization of novel phytocystatin gene from turmeric, *Curcuma longa*. BioMed Research International.2014.https://doi.org/10.1155/2014/973790

Chandhini, S. and Rejish Kumar, V. J. 2019.Transcriptomics in aquaculture: current status and applications. Reviews in Aquaculture.11(4):1379-1397.

Chandy, K.C., V.S. Pillai and P.K.V. Nambiar. 1980. Effect of ethyl methane sulfonate on the vegetative buds of pepper (*Piper nigrum* L.). Indian Cocoa, Arecanut, Spices J. 3:93-94.

Chang, K. L., and P. C. Ho. 2014. Gas chromatography time-of-flight mass spectrometry (GC-TOF-MS)-based metabolomics for comparison of caffeinated and decaffeinated coffee and its implications for Alzheimer's disease. PLoS One. 9. https://doi.org/10.1371/journal.pone.0104621

Chaudhary, R. and K.P.S. Chandel. 1994. Germination studies and cryopreservation of seeds of Black pepper (*Piper nigrum* L.), a recalcitrant species. Cryo-Letter. 15:145-150.

Chen, C., G. Chen, B. Cao and J. Lei. 2015b. Transcriptional profiling analysis of genic male sterile–fertile *Capsicum annuum*reveal candidate genes for pollen development and maturation by RNA-Seq technology. Plant Cell, Tissue and Organ Culture (PCTOC). 122:465-476.

Chen, H., L. Liu, L. Wang, S. Wang, P. Somta and X. Cheng. 2015a. Development and validation of EST-SSR markers from the transcriptome of adzuki bean (*Vigna angularis*). PloS One.10(7). https://doi.org/10.1371/journal.pone.0131939

Chen, R., H. Li, L. Zhang, J. Zhang, J. Xiao and Z. Ye. 2007. CaMi, a root-knot nematode resistance gene from hot pepper (*Capsium annuum* L.) confers nematode resistance in tomato. Plant Cell Reports. 26: 895-905.

Chiang, M. S., C. Chong, B.S. Landry and R. Crete. 1993. Cabbage: *Brassica oleracea* subsp. *capitata* L., p.113-155, In G. Kallo and B.O. Bergh, ed. Genetic improvement of vegetable crops. Pergamon.

Chombe, D., E. Bekele, T. Bryngelsson, A. Teshome and M. Geleta. 2017. Genetic structure and relationships within and between cultivated and wild korarima [*Aframomum corrorima* (Braun) PCM Jansen] in Ethiopia as revealed by simple sequence repeat (SSR) markers. BMC Genetics. 18: 72.

Choudhary, S., R. Sharma, G. Jethra, M.K. Vishal and A. Tripathi. 2019. Molecular diversity in coriander (*Coriandrum sativum*) using RAPD and ISSR markers. Indian Journal of Agricultural Sciences.89: 193-198.

Cocking, E. C. 1960.A method for the isolation of plant protoplasts and vacuoles. Nature.187: 962-963.

Collard, B. C. and D.J. Mackill. 2008. Marker-assisted selection: an approach for precision plant breeding in the twenty-first century. Philosophical Transactions of the Royal Society B: Biological Sciences. 363: 557-572.

Corinto, Gian Luigi (2014). Nikolai Vavilov's Centres of Origin of cultivated plants with a view to conserving agricultural biodiversity. Human Evolution, 29 (4): 285-301.

Costa, R., G. Pereira, I. Garrido, M.M. Tavares-de-Sousa and F. Espinosa. 2016. Comparison of RAPD, ISSR, and AFLP molecular markers to reveal and classify orchardgrass (*Dactylis glomerata* L.) germplasm variations. PloS One.11(4). https://doi.org/10.1371/journal.pone.0152972.

Criley, R. A., M.S. Roh, M. Kikuchi and R.M. Manshardt. 2008. A comparison of *Gardenia augusta* cultivars using isozymes and RAPD markers. Acta Horticulturae. 766:461-468.

Cruz-Cruz, C.A., M.T. González-Arnao and F. Engelmann. 2013. Biotechnology and conservation of plant biodiversity. Resources.2:73-95.

Cyriac, A., R. Paul, K. Anupama, T.E. Sheeja, K.N. Babu and V.A. Parthasarathy. 2016.Isolation and characterization of genomic microsatellite markers for small cardamom (*Elettaria cardamomum* Maton) for utility in genetic diversity analysis. Physiology and Molecular Biology of Plants. 22:219-229.

Dantu, P. K. and U.K. Tomar. 2010. Somatic Embryogenesis, p. 892-902, In G. Tripathi, ed. Cellular and Biochemical Science. IK International House Pvt Ltd, New Delhi.

Das kalov, S. 1986. Mutation breeding in pepper. Mutation Breeding review. 4: 26.

Das, A., V. Kesari, V.M. Satyanarayana, A. Parida and L. Rangan . 2011. Genetic relationship of *Curcuma species* from Northeast India using PCR-based markers. Molecular Biotechnology. 49:65-76.

Das, G., J.K. Patra and K.H. Baek. 2017. Insight into MAS: a molecular tool for development of stress resistant and quality of rice through gene stacking. Frontiers in Plant Science.8:985.

Dave, K. A., M.J. Headlam, T.P. Wallis and J.J. Gorman. 2011. Preparation and analysis of proteins and peptides using MALDI TOF/TOF mass spectrometry. Current Protocols in Protein Science.63:16-13.

Davey, M. R., P. Anthony, J.B. Power and K.C. Lowe. 2005. Plant protoplasts: status and biotechnological perspectives. Biotechnology Advances. 23:131-171.

David, D., T.Y. Ji and J.I. Gansau. 2016. *In vitro* propagation of *Zingiber officinale* Rosc.'Tambunan'. Transactions on Science and Technology.3:162-167.

Davies, P. J. 2010. The plant hormones: their nature, occurrence, and functions, p.1-15, In P. J. Davies, ed. Plant Hormones, Springer, Dordrecht.

De Figueiredo, R.A. and M. Sazima. 2000. Pollination biology of Piperaceae species in South eastern Brazil. Annals of Botany. 85:455–60.

De Waard, P.W.F., C.A. Zeven. 1969. Pepper, *Piper nigrum* L. In: Outlines of Perennial Crop Breeding in Tropics. Wageningen, The Netherlands. pp. 409–426.

Debergh, P. C. and L.J.Maene. 1981. A scheme for commercial propagation of ornamental plants by tissue culture. Scientia Horticulturae.14:335-345.

Debnath, M., B. De and S. Das. 2020. GC-MS-Based profiling of non-polar metabolites and chemometric study of fruits of *Capsicum* species and landraces at different stages of ripening. Journal of Herbs, Spices & Medicinal Plants.26:126-147.

Debnath, M., C.P.Malik and P.S. Bisen. 2006. Micropropagation: a tool for the production of high quality plant-based medicines. Current Pharmaceutical Biotechnology.7:33-49.

Dekkers, J. C. and F. Hospital.2002. The use of molecular genetics in the improvement of agricultural populations. Nature Reviews Genetics. 3: 22-32.

Dekkers, J.C.M. 2003. Commercial application of marker- and geneassisted selection in livestock: strategies and lessons. Presented at the 54 th annual meeting of the European Association for Animal Production, Rome, Italy, 31 August - 3 September.

Desmukh, M.V. 1982. Studies in the mutagenic effect of gamma rays and Ethyl Methane Sulphonate on chilli (*Capsicum annuum* L.) cv. Jwala. M.Sc. thesis, MPKV, Akola

Dhandapani, V., S.R. Choi, P. Paul, Y.K. Kim, N. Ramchiary, Y. Hur and Y.P. Lim. 2012. Development of EST database and transcriptome analysis in the leaves of Brassica rapa using a newly developed pipeline. Genes & Genomics. 34: 671-679.

Dhanya, K. and B. Sasikumar. 2010. Molecular marker based adulteration detection in traded food and agricultural commodities of plant origin with special reference to spices. Current Trends in Biotechnology and Pharmacy.4: 454-489.

Divakaran, M., G.S.Pillai, K.N. Babu and K.V. Peter. 2008.Isolation and fusion of protoplasts in Vanilla species. Current Science. 94:115-120.

Dokku, P., K.M. Das and G.J.N. Rao. 2013. Pyramiding of four resistance genes of bacterial blight in Tapaswini, an elite rice cultivar, through marker-assisted selection. Euphytica.192: 87-96.

Domon, B. and R. Aebersold. 2006. Mass spectrometry and protein analysis. Science.312: 212-217.

Dong, Z. and Y. Chen. 2013 .Transcriptomics: advances and approaches. Science China Life Sciences.56: 960-967.

Diederichsen, A. 1996. Coriander (*Coriandrum sativum* L.): promoting the conservation and use of underutilized and neglected crops. 3. Institute of Plant Genetics and Crop Plant Research, Gatersleben/International Plant Genetic Resources Institute, Rome.

Dunwell, J. M. 2010. Haploids in flowering plants: origins and exploitation. Plant Biotechnology Journal.8:377-424.

Durieu, P. and S.J. Ochatt. 2000. Efficient intergeneric fusion of pea (*Pisum sativum* L.) and grass pea (*Lathyrus sativus* L.) protoplasts. Journal of Experimental Botany.51:1237-1242.

Ebrahimi, M., A. Mokhtari and R. Amirian. 2018. A highly efficient method for somatic embryogenesis of *Kelussia odorotissima* Mozaff., an endangered medicinal plant. Plant Cell, Tissue and Organ Culture (PCTOC).132: 99-110.

Egan, A. N., J. Schlueter and D.M. Spooner. 2012. Applications of next generation sequencing in plant Biology. American Journal of Botany. 99(2): 175-185.

Elangbam, M. and A.K. Misra. 2016. Development of CAPS markers to identify Indian tea (*Camellia sinensis*) clones with high catechin content. Genetics and Molecular Research. 15:1-13.

El-Mahrouk, M. E., M.K. Maamoun, A.N. El-Banna, S.A. Omran, Y.H. Dewir and S. El-Hendawy. 2018. *In vitro* gynogenesis and flow cytometry analysis of the regenerated haploids of black cumin (*Nigella sativa*). HortScience.53:681-686.

Emons, A. M. C. 1994. Somatic embryogenesis: cell biological aspects. Acta Botanica Neerlandica. 43:1-14.

Engelmann, F. 1991. *In vitro* conservation of tropical plant germplasm—a review. Euphytica. 57: 227-243.

Engelmann, F. 2000.Importance of cryopreservation for the conservation of plant genetic resources. p.8-20, In Engelmann, F. and H. Takagi, ed. Proceedings of an international workshop, Tsukuba, Japan, October, 1998.

Engelmann, F. 2011.Use of biotechnologies for the conservation of plant biodiversity. *In Vitro* Cellular & Developmental Biology Plant. 47:5-16.

Escudero, N., F.Marhuenda-Egea and L. V. Lopez-Llorca. 2017. Metabolomics. p.169-181, In R.. H. Manzanilla-López and L.V. Lopez-Llorca, ed.Perspectives in Sustainable Nematode Management Through Pochoniachlamydosporia Applications for Root and Rhizosphere Health.Springer, Cham.

Espinosa-Leal, C. A., C.A. Puente-Garza and S. García-Lara. 2018. *In vitro* plant tissue culture: means for production of biological active compounds. Planta.248:1-18.

Esselman, E. J., L. Jianqiang, D.J. Crawford, J.L. Windus and A.D. Wolfe. 1999. Clonal diversity in the rare *Calamagrostis porterissp. insperata* (Poaceae): comparative results for allozymes and random amplified polymorphic DNA (RAPD) and inter simple sequence repeat (ISSR) markers. Molecular Ecology.8: 443-451.

Everley, R.A., T.M. Mott, S.A. Wyatt, D.M. Toney and T.R. Croley. 2008. Liquid chromatography/mass spectrometry characterization of *Escherichia coli* and *Shigella species*. Journal of the American Society for Mass Spectrometry.19:1621-1628.

Fahy, G. M., D.R. MacFarlane, C.A. Angell and H.T. Meryman. 1984. Vitrification as an approach to cryopreservation. Cryobiology. 21: 407-426.

Falvo, S., M. Di Carli, A. Desiderio, E. Benvenuto, A. Moglia, T. America, ...and A. Acquadro. 2012. 2-D DIGE analysis of UV-C radiation responsive proteins in globe artichoke leaves. Proteomics.12:448-460.

Farag, M. A., H.A. Gad, A.G. Heiss and L.A. Wessjohann. 2014. Metabolomics driven analysis of six Nigella species seeds via UPLC-qTOF-MS and GC–MS coupled to chemometrics. Food Chemistry.151:333-342.

Farroqui, A.A. and B.S.Sreeramu. 2005. Tropical Spice Crops and their cultivation, University Press, Hyderabad.

Feng, S., Y, Zhu, C. Yu, K. Jiao, M. Jiang,... and H. Wang. 2018. Development of species-specific SCAR markers, based on a SCoT analysis, to authenticate *Physalis* (Solanaceae) *species*. Frontiers in Genetics. 9:192.https://doi.org/10.3389/fgene.2018.00192

Ferrie, A. M. R. and K.L. Caswell. 2011. Isolated microspore culture techniques and recent progress for haploid and doubled haploid plant production. Plant Cell, Tissue and Organ Culture (PCTOC).104:301-309.

Finehout, E. J. and K.H. Lee. 2004. An introduction to mass spectrometry applications in biological research. Biochemistry and Molecular Biology Education.32(2):93-100.

Firoozi, B., N. Zare, O. Sofalian and P.S. Mosadegh. 2018. *In vitro* indirect somatic embryogenesis and secondary metabolites production in the saffron: emphasis on ultrasound and plant growth regulators. Journal of Agricultural Sciences.25:1-10.

Fischer, C. and G.Neuhaus. 1996. Influence of auxin on the establishment of bilateral symmetry in monocots. The plant Journal.9:659-669.

Fonseka, D. and W.Wickramaarachchi. 2018. *In vitro* shoot regeneration and rooting of *Piper Longum* L.: A valuable Medicinal Plant. International Journal for Research in Applied Sciences and Biotechnology.5:10-13.

Francia, E., G. Tacconi, C. Crosatti, D. Barabaschi, D. Bulgarelli, E. Dall'Aglio and G. Valè. 2005. Marker assisted selection in crop plants. Plant Cell, Tissue and Organ Culture (PCTOC). 82: 317-342.

Frisch, M., M. Bohn and A.A. Melchinger. 1999. Minimum sample size and optimal positioning of flanking markers in marker assisted backcrossing for transfer of a target gene. Crop Science.39:967-975.

Fryer, R. M., J. Randall, T. Yoshida, L. L. Hsiao, J. Blumenstock, K.E. Jensen, ...and S.R. Gullans, S. R. 2002. Global analysis of gene expression: methods, interpretation, and pitfalls. Nephron Experimental Nephrology.10: 64-74.

Gad, H. A. and A. Bouzabata.2017. Application of chemometrics in quality control of Turmeric (*Curcuma longa*) based on Ultra-violet, Fourier transform-infrared and 1H NMR spectroscopy. Food chemistry.237:857-864.

Gaj, M.D. 2004. Factors influencing somatic embryogenesis induction and plant regeneration with particular reference to *Arabidopsis thaliana* (L.) Heynh. Plant Growth Regulation.43:27-47.

Ganga. G., V.R. Sandeep and K.G. Johnson. 2004. ISSR and RAPD markers in diversity analysis of Piper species. p.639. In: Proc. Sixteenth Kerala Science Congress, January 29-31, Kozhikode, India.

Ganie, S. H., P. Upadhyay, S. Das and M.P. Sharma. 2015. Authentication of medicinal plants by DNA markers. Plant Gene. 4: 83-99.

Gantait, S., N. Mondal and S. Nandi.2011. Advances in micropropagation of selected aromatic plants: a review on vanilla and strawberry. American Journal of Biochemistry and Molecular Biology.1:1-19.

Garg, R., R.K. Patel, S. Jhanwar, P.Priya, A. Bhattacharjee, G. Yadav and M. Jain. 2011. Gene discovery and tissue-specific transcriptome analysis in chickpea with massively parallel pyrosequencing and web resource development. Plant Physiology. 156: 1661-1678.

Geetha, S.P., C. Manjula and A. Sajina. 1995. *In vitro* conservation of genetic resources of spices. In. Proc. Seventh Kerala Science congress, January 27-29, STEC, Kerala, pp.12-16.

Geetha, S.P., K.N. Babu, J. Rema, P.N. Ravindran and K.V. Peter. 2000. Isolation of protoplasts from cardamom (*Elettaria cardamomum* Maton.) and ginger (*Zingiber officinale* Rosc.). Journal of Spices and Aromatic Crops. 9 : 23-30.

Gentry, H.S. 1955.Apomixis in black pepper and Jojoba. Journal of Heredity. 46:8–13 .

Germana'. M. A. 2011. Gametic embryogenesis and haploid technology as valuable support to plant breeding. Plant Cell Reports.30:839-857.

Ghanti, K., C.P. Kaviraj, R.B. Venugopal, F.T.Z. Jabeen and S. Rao. 2004. Rapid regeneration of *Mentha piperita* L. from shoot tip and nodal explants. Indian Journal Biotechnology. 3: 594-598.

Giachino, R. R. A. 2020. Investigation of the genetic variation of anise (*Pimpinella anisum* L.) using RAPD and ISSR markers. Genetic Resources and Crop Evolution. 67: 763-780.

Giaever, K. O., H. Edvard and F. Bjørndalen. 2017. A study of Marker assisted backcrossing (MAB) benefits of molecular markers in breeding programs. International Journal of Animal Breeding and Genetics.1:001-016.

Gill, K., A.K. Singh, S. Kumar, B. Mishra, V. Kapoor, S.N. Das and S. Dey. 2012. Isolation and characterization of a potent protein from ginger rhizomes having multiple medicinal properties. Research Journal of Medicinal Plant.6:160-170.

Girija, D., P.S. Beena and P.A. Nazeem. 2005a. Molecular cloning of a cDNA fragment encoding the defense related protein β-1,3-glucanase in black pepper (*P. nigrum* L.) Proc. Kerala Science Congress, January 2005, KFRI, Peechi, Kerala pp. 81-82.

Girija, D., P.S. Beena, P.A. Nazeem and M.G. Puroshothama. 2005b. Molecular cloning of cDNA fragment encoding hydroxy methyl glutaryl CoA reductase in *Piper colubrinum.* Proc. National symposium on biotechnological interventions for improvement of horticultural crops: issues and strategies. 10-12 January, 2005, Kerala Agricultural University, Thrissur, Kerala, pp:303-306

Gomes, H.T., A. Martins, S.O.D. de Oliveira and J.E. Scherwinski-Pereira. 2015 Assessment of mint (*Mentha* spp.) species for large-scale production of plantlets by micropropagation. Embrapa Recursos Genéticos e Biotecnologia-Artigo em periódico indexado (ALICE). 37:405-410.

Gondane, B.G., D.T. Deshmukh. 2004. Exploitation of heterosis in chilli. J Soils Crops. 14:376-82 .

Goodacre, R., S. Vaidyanathan, W.B. Dunn, G.G. Harrigan and D.B. Kell. 2004. Metabolomics by numbers: acquiring and understanding global metabolite data. Trends in Biotechnology. 22: 245-252.

Govindaraj, M., M. Vetriventhan and M. Srinivasan, M. 2015. Importance of genetic diversity assessment in crop plants and its recent advances: an overview of its analytical perspectives. Genetics Research International.2015.https://doi.org/10.1155/2015/431487

Gray, D. J., M. E. Compton, R.C. Harrell and D. J. Cantliffe. 1995. Somatic embryogenesis and the technology of synthetic seed., In: Y.P.S.Bajaj, ed. Somatic Embryogenesis and Synthetic Seed I (Part of Biotechnology in Agricultural Forestry Series, Volume.60), Springer, Berlin, Heidelberg. pp.126-151.

Gregory, P.J. 1936. The floral morphology and cytology of *Elettaria cardamomum* Maton). J. Linn. Soc. (Lond) 50:363-393.

Grover, S. and C.P. Malik. 2017. Genetic diversity and identification of variety-specific RAPD and ISSR markers in *Foeniculum vulgare*. LS: International Journal of Life Sciences. 6:31-38.

Guan, Q., Y. Guo, Y. Wei, F. Meng and Z. Zhang. 2010. Regeneration of somatic hybrids of ginger via chemical protoplast fusion. Plant Cell, Tissue and Organ Culture (PCTOC).102:279-284.

Guha, S. and S.C. Maheshwari. 1964. *In vitro* production of embryos from anthers of Datura. Nature.204:497.

Guijarro-Díez, M., L. Nozal, M.L. Marina and A.L. Crego. 2015. Metabolomic fingerprinting of saffron by LC/MS: novel authenticity markers. Analytical and Bioanalytical Chemistry. 407: 7197-7213.

Guo, D.L., J.Y. Zhang and C.H. Liu. 2012. Genetic diversity in some grape varieties revealed by SCoT analyses. Molecular Biology Reports.39: 5307-5313.

Guo, G., G. Zhang, B. Pan, W. Diao, J. Liu, W. Ge, and S. Wang. 2019. Development and application of InDel markers for *Capsicum* spp. based on whole-genome re-sequencing. Scientific Reports. 9(1):3691.

Guo, W.L., R.G. Chen, X.H. Du, Z. Zhang, Y.X. Yin, Z.H. Gong and G.Y. Wang. 2014. Reduced tolerance to abiotic stress in transgenic Arabidopsis overexpressing a Capsicum annuummultiprotein bridging factor 1. BMC Plant Biology. 14:138.

Guo, Y., J. Bai and Z. Zhang. 2007. Plant regeneration from embryogenic suspension-derived protoplasts of ginger (*Zingiber officinale* Rosc.). Plant Cell, Tissue and Organ Culture (PCTOC).89:151-157.

Gupta, S. M., A. Grover and M. Nasim. 2013. Transgenic technologies in agriculture: from lab to field to market. CIBTech Journal of Biotechnology. 3(3):20-47.

Gupta, V., P.K. Jatav, S.U. Haq, K. S. Verma, V. K. Kaul, S.L Kothari and S. Kachhwaha. 2019 . Translation initiation codon (ATG) or SCoT markers-based polymorphism study within and across various Capsicum accessions: insight from their amplification, cross-transferability and genetic diversity. Journal of Genetics. 98: 61.

Guzmán, F. A., S. Moore, M.C. de Vicente and M. M. Jahn. 2020. Microsatellites to enhance characterization, conservation and breeding value of *Capsicum* germplasm. Genetic Resources and Crop Evolution.67:569-585.

Hadian, J., A. Karami, A. Azizi and A. Khadivi-Khub. 2015. Ubiquitous genetic diversity among and within wild populations of *Satureja rechingeri* assessed with ISSR markers. Plant Systematics and Evolution.301:923-930.

Hamirah, M. N., H.B. Sani, P.C. Boyce and S.L. Sim. 2010. Micropropagation of red ginger (*Zingiber montanum* Koenig), a medicinal plant. *Asia Pacific Journal of Molecular Biology* and *Biotechnology.* 18(1):127-130.

Han, J. S., S. Lee, H.Y. Kim and C.H. Lee. 2015. MS-based metabolite profiling of aboveground and root components of *Zingiber mioga* and *Z. officinale*. Molecules. 20(9):16170-16185.

Hasan, M. M., M.Y. Rafii, M.R. Ismail, M. Mahmood, H. A. Rahim, M.A. Alam, and M.A. Latif. 2015. Marker-assisted backcrossing: a useful method for rice improvement. Biotechnology & Biotechnological Equipment. 29: 237-254.

Haynes, P. A. and J.R. Yates III. 2000. Proteome profiling—pitfalls and progress. Yeast.17: 81-87.

Hegde, V., P.S. Partap and R.C. Yadav .2017a. .*In vitro* regeneration of Capsicum (*Capsicum annuum* L.) from cotyledon Explants. International Journal of Current Microbiology and Applied Sciences.6:225-237.

Hegde, V., P.S. Partap, R.C. Yadav and K.S. Baswana. 2017b. *In vitro* Androgenesis in Capsicum (*Capsicum annuum* L.). International Journal of Current Microbiology and Applied Sciences. 6(5):925-933.

Herison, C., Rustikawat and Sudarsono. 2004. Genetic nature of resistance against. Cucumber Mosaic Virus in hot pepper. Capsicum and Eggplant Newsletter.23:111-114.

Hesse, H. and R. Höfgen. 2001. Application of genomics in agriculture.p.61-79 M.J. Hawkesford and P.Buchner, eds. In Molecular analysis of Plant Adaptation to the Environment.Springer, Dordrecht.

Hofmann, A., J. Proust, A. Dorowski, R. Schantz and R. Huber. 2000. Annexin 24 from *Capsicum annuum* X-ray structure and biochemical characterization. Journal of Biological Chemistry.275:8072-8082.

Horgon, R. P and L. C. Kenny. 2011. SAC review: Omic technologies: genomics, transcriptomics. Obstetrician Gynaecologist. 13:189–195.

Hospital, F., L. Moreau, F. Lacoudre, A. Charcosset and A. Gallais. 1997. More on the efficiency of marker-assisted selection. Theoretical and Applied Genetics.95: 1181-1189.

Hossein, S., R. Ibrahim and P. K.A. Ling. 2006. Somatic embryogenesis: an alternative method for *in vitro* micropropagation.Iranian Journal of Biotechnology. 4:156-161.

http://agriinfo.in/default.aspx?page=topicandsuperid=2andtopicid=1393

http://shodhganga.inflibnet.ac.in/bitstream/10603/3949/8/08_chapter%201.pdf

http://www.cabi.org/isc/datasheet/13573

http://www.worldagroforestry.org/treedb/AFTPDFS/Cinnamomum_verum.PDF

https://www.ncbi.nlm.nih.gov/probe/docs/techrapd/

https://www.ncbi.nlm.nih.gov/probe/docs/techrapd/

Hu, L., C. Hao, R. Fan, B. Wu, L. Tan and H. Wu. 2015. De novo assembly and characterization of fruit transcriptome in black pepper (*Piper nigrum*). Plos One.10. https://doi.org/10.1371/journal.pone.0136028

Huaizhu, L., Y. Lin, J. Bai, A. Qiuyan, D. Lingling and S. Rui-Xue. 2020. The complete chloroplast genome sequence of *Mentha canadensis* (Labiatae), a traditional Chinese herbal medicine. Mitochondrial DNA Part B. 5:55-56.

Hunault, G., A.Maatar. 1995. Enhancement of somatic embryogenesis frequency by gibberellic acid in fennel. Plant Cell Tissue and Organ Culture (PCTOC). 41: 171-176.

Hussain, A., I.A. Qarshi, H. Nazir and I. Ullah. 2012. Plant tissue culture: current status and opportunities, p.1-28, In A. Leeva, ed. Recent Advances in Plant *in vitro* Culture, IntechOpen.

Hussain, A., S. Naz, H. Nazir and Z.K. Shinwari. 2011.Tissue culture of black pepper (*Piper nigrum* L.) in Pakistan. Pakistan Journal of Botany. 43(2):1069-1078.

Ibitoye, D. O. and P.E. Akin-Idowu. 2011. Marker-assisted-selection (MAS): A fast track to increase genetic gain in horticultural crop breeding. African Journal of Biotechnology. 10: 11333-11339.

Ibrahim, D. A., G.H. Danial, V.M. Mosa and B.M. Khalil. 2015. Plant regeneration from shoot tips-derived callus of ginger (*Zingiber officinale* Rosc.). Journal of Experimental Agriculture International.7: 55-61.

Ibrahim, K.K., V.S. Pillai and S. Sasikumaran. 1987b. An extra bold natural mutant of Panniyur -1 variety of black pepper. Indian Cocoa Arecanut and Spice J. 21: 58.

Igwe, D. O., C.A. Afiukwa, G. Acquaah and G.N. Ude. 2019. Genetic diversity and structure of *Capsicum annuum* as revealed by start codon targeted and directed amplified minisatellite DNA markers. Hereditas.156: 32. https://doi.org/10.1186/s41065-019-0108-6

IISR. 2004. Annual Report 2003-2004, Indian Institute of Spices Research, Calicut, India.

Ilahi, I. and M. Jabeen. 1992. Tissue culture studies for micropropagation and extraction of essential oils from *Zingiber officinale* Rosc. Pakistan Journal of Botany. 24:54-54.

Imadi, S. R., A.G. Kazi, M.A. Ahanger, S. Gucel and P. Ahmad. 2015. Plant transcriptomics and responses to environmental stress: an overview. Journal of Genetics.94:525-537.

IPGRI, 1994. Descriptor for cardamom (Elettaria cardamomum Maton.).International Plant Genetic Resources Institute, Rome, pp 27-41. https://www.bioversityinternational.org/e-library/publications/detail/descriptors-for-cardamom-elettaria-cardamomum-maton/

Irulappan, I. V Ponnuswamy, E. Vadivel and G. Dharmaraj.1982. Sensitivity studies in Pepper (*Piper nigrum* L.) South Indian Hort. 30:51-53.

Islam, A. R., M.M. Islam and M.F. Alam. 2017. Rapid *in vitro* clonal propagation of herbal spice, *Mentha piperita* L. using shoot tip and nodal explants. Science and Education.5:43-50.

Islam, A. T. M. R. and M.F. Alam. 2018. *In vitro* callus induction and indirect organogenesis of *Mentha piperita* (L.)-an aromatic medicinal plant. GSC Biological and Pharmaceutical Sciences.4:49-60.

Islam, M. A., A. Meister, V. Schubert, K. Kloppstech and E. Esch. 2007. Genetic diversity and cytogenetic analyses in *Curcuma zedoaria* (Christm.) Roscoe from Bangladesh. Genetic Resources and Crop Evolution.54: 149-156.

Islek, C., B.T. Unal, E. Koc and D. Kaya. 2016. Effect of Thidiazuron used as an elicitor in the production of capsaicin on total protein and phenolic amounts, antioxidant enzyme activities of pepper plants. International Journal of Environment, Agriculture and Biotechnology (IJEAB).1(3):603-609.

Jacobs, D. I., R. van der Heijden and R. Verpoorte. 2000. Proteomics in plant biotechnology and secondary metabolism research. Phytochemical Analysis: An International Journal of Plant Chemical and Biochemical Techniques. 11(5):277-287.

Jaiswal, Y., Z. Liang, A. Ho, H. Chen and Z. Zhao. 2014. Tissue-Specific Metabolite Profiling of Turmeric by Using Laser Microdissection, Ultra-High Performance Liquid Chromatography-Quadrupole Time of Fight-Mass Spectrometry and Liquid Chromatography-Tandem Mass Spectrometry. European Journal of Mass Spectrometry. 20:383-393.

Jang, Y. K., E.S. Jung, H.A. Lee, D. Choi and C.H. Lee. 2015.Metabolomic characterization of hot pepper (*Capsicum annuum* "CM334") during fruit development. Journal of Agricultural and Food chemistry.63:9452-9460.

Jayakumar, V.N., K. Nirmal Babu and M.N. Venugopal.2005. Molecular characterization of small cardamom (ElettariacardamomumMaton.) collections and related genera using RAPD, PCR -RFLP and ISSR markers. In. Abst. National Symposium on Biotechnological interventions for improvement of Horticultural Crops: Issues and Strategies, 10- 12 January 2005, Kerala Agricultural University, Trissur, Kerala, India: pp.122.

Jayaram, N., A.M. Rao, S. Ramesh, B. Manjunath, N. Mangalagowri, and M. Ashwini. 2016. . Tagging SSR markers to genotic regions associated with anthracnose disease resistance and productivity per se traits in Hot Pepper (*Capsicum annuum* L.). Environment and Ecology.34 : 1440-1446.

Jazayeri, S. M and R.O.V. Torres. 2017. Genomic and transcriptomic approaches toward plant selection. Journal of Science and Research: RevistaCiencia e Investigación. 2: 54-64.

Jeong, B.R., K.Fujiwara and T. Kozai. 1995.Environmental control and photoautotrophic micropropagation. Horticultural Reviews.17:125–172.

Jeong, K., D. Choi and J. Lee. 2018. Fine mapping of the genic male-sterile ms 1 gene in *Capsicum annuum* L. Theoretical and Applied Genetics.131: 183-191.

Jethra, G., S. Choudhary and V. Sharma. 2017. Development of new set of EST-SSR markers for evaluating genetic diversity and fingerprinting Local Celery (*Apium graveolens* L.) Cultivars. Current Biomarkers (Formerly: Recent Patents on Biomarkers).7: 47-57.

Jiang, C., L. Cao, Y. Yuan, M. Chen, Y. Jin and L. Huang. 2014. Barcoding melting curve analysis for rapid, sensitive, and discriminating authentication of saffron (*Crocus sativus* L.) from its adulterants. BioMed Research International. 2014.https://doi.org/10.1155/2014/809037

Jiang, G. L. 2015. Molecular marker-assisted breeding: a plant breeder's review. p.431-472, In J. M. Al-Khayri, S.M. Jain and D. V. Johnson, eds. Advances In Plant Breeding Strategies: Breeding, Biotechnology And Molecular Tools. Springer,Cham.

Jiang, G. L.2013. Molecular markers and marker-assisted breeding in plants. In: S.V.Andersen, ed. Plant Breeding from Laboratories to Fields, InTech, Croatia. pp.45-83.

Jie, E. Y., M.S. Ahn, J. Lee, Y.I. Cheon, C.Y. Kim and S.W. Kim. 2019. Establishment of a high-frequency plant-regeneration system from rhizome-derived embryogenic cell-suspension cultures of *Curcuma longa* L. Plant Biotechnology Reports.13:123-129.

Jiménez, V. M. 2001. Regulation of *in vitro* somatic embryogenesis with emphasis on to the role of endogenous hormones. Revista Brasileira de Fisiologia Vegetal.13 : 196-223.

Jiménez, V. M. 2005. Involvement of plant hormones and plant growth regulators on *in vitro* somatic embryogenesis. Plant Growth Regulation.47:91-110.

Jin, S., C. Song, S. Jia, S. Li, Y. Zhang, C. Chen and H. Jiang.2017. An integrated strategy for establishment of curcuminoid profile in turmeric using two LC–MS/MS platforms. Journal of Pharmaceutical and Biomedical Analysis.132:93-102.

Jogarana, M. R., R.V. Bhadani, H.J. Kachhadiya, A.K. Nandha, M. Kapuria, A. Virali and R.S. Tomar. 2019. Development and validation of simple sequence repeat markers from genome of ajwain (*Trachyspermum ammi* L.). International Journal of Chemical Studies.7: 426-432.

Johnson, G.K., G. Ganga, S. Varma *et al.* 2005. Identification of hybrids in black pepper (*Piper nigrum* L.) using male parent-specific RAPD markers. Current Science. 88:1-2.

Johnson, G.K., S. Varma, G. Ganga *et al.* 2003. ISSR-PCR, a potential tool for genetic diversity analysis in spices. In: Proc National seminar on new perspectives in Spices, Medicinal and Aromatic Plants, 27-29 November, Goa, pp. 23-26.

Johri, J.K., Aminuddin and P. Aruna. 1996. Regeneration of betelvine (*Piper betle* L.) through somatic embryogenesis. Indian Journal of Experimental Biology. 34:83-85.

Jose, S., R. Sujatha and K.P. Deeshma. 2018. Novel EST-SSR marker development and validation in Black Pepper cultivars and varieties. Journal of Tropical Agriculture.55:175-179.

Joseph, B., D. Joseph, S. Jose, C.A. Mathews and M.M.Mathew. 2012. *In vitro* studies on five varieties of Black Pepper. Journal of Tropical Medicinal Plants. 13: 145-151

Joseph, L., P.A. Nazeem, S.T. Mini, S. Philip and M. Balachandran. 1996. *In vitro* techniques for mass multiplication of black pepper (*Piper nigrum* L.) and ex vitro performance of the plantlets. J. Plantation Crops (Suppl.) 24: 511-516.

Joshi, S. P., P.K. Ranjekar and V.S. Gupta. 1999. Molecular markers in plant genome analysis. Current Science. 77: 230-240.

Kambaska, K. B. and S.Santilata. 2009. Effect of plant growth regulator on micropropagation of ginger (*Zingiber officinale* Rosc.) cv-Suprava and Suruchi. Journal of Agricultural Technology. 5:271-280.

Kanthaswamy, V., K.Hemavathy, D.Veeragavathatham and K.Srinivasan. 2003. Effect of physiological basis on heterosis of chilli (*Capsicum annuum* L.). South Indian Hort. 51: 157-162.

Kao, K. N. and M. Michayluk. 1974. A method for high-frequency intergeneric fusion of plant protoplasts. Planta.115:355-367.

Kar, B., S. Nayak and R.K. Joshi.2014b. Development and evaluation of STS diagnostic marker to track turmeric (*Curcuma longa* L.) resistance against rhizome rot caused by *Pythium aphanidermatum*. Australasian Plant Pathology.43:167-175.

Kar, D., P.K. Pattanaik, L. Acharya, M.K. Panda, K. Sathapathy, A. Kuanar and B. Mishra. 2014a. Assessment of genetic diversity among some elite cultivars of ginger (*Zingiber officinale* Rosc.) using isozyme and protein markers. Brazilian Journal of Botany. 37: 469-479.

Karuppusamy, S. 2009. A review on trends in production of secondary metabolites from higher plants by *in vitro* tissue, organ and cell cultures. Journal of Medical Plants Research.3:1222-1239.

KAU, 2001. Three decades of spices research at KAU. Directorate of extension, Kerala Agricultural University, Thrissur, Kerala, India.pp: 37-55.

Kesawat, M. S. and B.D. Kumar. 2009. Molecular markers: it's application in crop improvement. Journal of Crop Science and Biotechnology.12: 169-181.

Keshavachandran, R., P.A. Nazeem and J.L. Karihaloo. 2005. Genetic fingerprinting of *Piper nigrum* L. and *Piper longum* L. Cultivars using RAPD markers. Proceedings of ICAR National Symposium on Biotechnological interventions for improvement of Horticultural Crops: Issues and Strategies, Kerala Agricultural University, Trissur, Kerala, India. pp.288 – 290.

Keskitalo, M., P. Angers, E. Earle and E. Pehu. 1999. Chemical and genetic characterization of calli derived from somatic hybridization between tansy (*Tanacetum vulgare* L.) and pyrethrum (*Tanacetum cinerariifolium* (Trevir.) Schultz-Bip.). Theoretical and Applied Genetics.98:1335-1343.

Key, S., J.K. Ma and P.M. Drake. 2008. Genetically modified plants and human health. Journal of the Royal Society of Medicine.101:290-298.

Khan, E. U. and J.H.Liu. 2009. Plant biotechnological approaches for the production and commercialization of transgenic crops. Biotechnology & Biotechnological Equipment.23: 1281-1288.

Khlestkina, E. K. and E.A. Salina. 2006. SNP markers: methods of analysis, ways of development, and comparison on an example of common wheat. Russian Journal of Genetics. 42: 585-594.

Khumaida, N., S.W. Ardie, A. Setiadi and L.N. Artiningsih. 2019. *In vitro* multiplication and acclimatization of black galingale (*Curcuma Aeruginosa* Roxb.). Journal of Applied Pharmaceutical Science. 9:110–116.

Kim, S., M. Park, S.I. Yeom, Y.M. Kim, J.M. Lee, H.A. Lee, and K. Jung. 2014. Genome sequence of the hot pepper provides insights into the evolution of pungency in *Capsicum* species. Nature Genetics.46:270-278.

Kiran, U., S. Khan, K.J. Mirza, M. Ram and M.Z. Abdin, M. Z. 2010. SCAR markers: a potential tool for authentication of herbal drugs. Fitoterapia.81.969-976.

Kishore, K., D. Rinchen, and B.Pandey. 2012. Floral boilogy of large cardamom (*Amomum subulatum*). Indian Journal of Agricultural Sciences. 82. 578-582.

Komamine, A., N. Murata and K. Nomura. 2005. Mechanisms of somatic embryogenesis in carrot suspension cultures-morphology, physiology, biochemistry, and molecular biology. *In Vitro* Cell Developmental Biology Plant.41:6-10.

Kong, X. M., Q. Zhou, F. Luo, B.D. Wei, Y.J. Wang, H.J. Sun, and S.J. Ji. 2019. Transcriptome analysis of harvested bell peppers (*Capsicum annuum* L.) in response to cold stress. Plant Physiology and Biochemistry.139: 314-324.

Koo, H., E.T. McDowell, X. Ma, K.A. Greer, J. Kapteyn, Z. Xie and D.R. Gang. 2013. Ginger and turmeric expressed sequence tags identify signature genes for rhizome identity and development and the biosynthesis of curcuminoids, gingerols and terpenoids. BMC Plant Biology.13: 27. http://www.biomedcentral.com/1471-2229/13/27

Kordrostami, M. and M. Rahimi. 2015. Molecular markers in plants: concepts and applications. *Genetics in The 3rd Millennium. 13*:4024-4031.

Krishnamoorthy, B. and V.A. Parthasarathy. 2009. Improvement of black pepperPerspectives in Agriculture, Veterinary Science, Nutrition and Natural Resources. 4: 85.

Krishnamurthy K S, Biju C N, Prasath D, Senthil Kumar C M, Kandiannan K, Maiti C S, Pauline Alila, Akali Sema, Aviboli Zhimomi & Nirmal Babu K (Eds) 2018. Souvenir and Abstracts, National Symposium on Spices and Aromatic Crops (SYMSAC IX): Spices for Doubling Farmer's Income, Indian Society for Spices, Kozhikode, Kerala, IndiaKrishnamoorthy, B. and V.A. Parthasarathy. 2010. Improvement of black pepper. Perspectives in Agriculture. Veterinary Science Nutrition and Natural Resources. 5: 1-12.

Krishnamoorthy, B., J.Remain, and P.A.Mathew. 2005. Biodiversity in spices. Spice India, 18:39-41.

Krishnamurthy, K., M.M. Khan, K.K. Avadhani, J. Venkatesh, A.L. Siddaramaiah, A.K. Chakravarthy and B.R.Gurumurthy.1989. Three decades of Cardamom Research at Regional Research Station, Mudigere, UAS, Bangalore 97.

Kumar, K.B., P. Kumar, S.M.Balachandran and R.D.Iyer.1985. Development of clonal plantlets from immature panicles of cardamom. Journal of Plantation Crops. 13. 31-34.

Kumar, M., V.R. Sharma, V. Kumar, U. Sirohi, V. Chaudhary, S.Sharma, ... and S. Sharma, S. 2019. Genetic diversity and population structure analysis of Indian garlic (*Allium sativum* L.) collection using SSR markers. Physiology and Molecular Biology of Plants.25: 377-386.

Kumar, O. A., S.S. Tata and T. Rupavathi.2014 .Evaluation of genetic diversity in 21 cultivars of chili pepper (*Capsicum annuum* L.) using isozyme markers. European Journal of Experimental Biology.4: 44-49.

Kumar, P., V.K. Gupta, A.K. Misra, D.R. Modi and B.K. Pandey. 2009. Potential of molecular markers in plant biotechnology. Plant Omics.2:141-162.

Kumar, S., S. Kumar, M. Singh, A. K. Singh and M.Rai. 2006. Identification of host plant resistance to pepper leaf curl virus in chilli (*Capsicum* sp.). Scientia Horticulturae.110: 359– 61.

Kumari, R., D. P. Wankhede, A. Bajpai, A. Maurya, K. Prasad, D. Gautam, D., ... and A.B. Gaikwad. 2019. Genome wide identification and characterization of microsatellite markers in black pepper (*Piper nigrum*): A valuable resource for boosting genomics applications. Plos One.14.https://doi.org/10.1371/journal.pone.0226002

Kumla, S., S. Doolgindachbaporn, R. Sudmoon and N. Sattayasai. 2012. Genetic variation, population structure and identification of yellow catfish, *Mystus nemurus* (C&V) in Thailand using RAPD, ISSR and SCAR marker. Molecular Biology Reports.39: 5201-5210.

Kuruvilla, K.M., K.J. Madhusoodanan, M.R.Sudharshan, P. Natarajan and J.Thomas. 2005. Performance evaluation of tissue culture Vs open pollinated seedling of cardamom. In: Proc. ICAR National Symposium on Biotechnological interventions for improvement of Horticultural Crops: Issues and Strategies, Kerala Agricultural University, Trissur, Kerala, India. pp. 81-83.

Kwon, S. J., S.K. Roy, J.H. Yu, S.W. Cho, H.H. Kim, H.O. Boo and S.H. 2019. Protein Profiling from Hormone-Induced Tetraploid Roots in *Platycodon grandiflorum*. Journal of Crop Science and Biotechnology.22: 465-474.

Lage, M. and C.L. Cantrell. 2009. Quantification of saffron (*Crocus sativus* L.) metabolites crocins, picrocrocin and safranal for quality determination of the spice grown under different environmental Moroccan conditions. Scientia Horticulturae.121:366-373.

Lal, R. K., S. P. S. Khanuja and H.O. Misra.1996. Genetic diversity in fennel (Foeniculum vulgare Miller). Indian Journal of Genetics and Plant Breeding. 66: 65-66 .

Lanigan, T.M., H.C. Kopera and T.L. Saunders. 2020. Principles of Genetic Engineering. Genes. 11: 291.Doi: 10.3390/genes11030291

Lebot V., K.M. Aradhya and R.M. Manshardt 1991. Geographical survey of genetic variation in Kava (*Piper methysticum* Forst. f. and *P. wichmannii* C. DC.). Pacific Science. 45: 169–187.

Lee, J., J.B. Yoon, J.H. Han, W.P. Lee, J.W. Do, H. Ryu and H. G. Park. 2010. A codominant SCAR marker linked to the genic male sterility gene (ms1) in chili pepper (*Capsicum annuum*). Plant Breeding.129: 35-38.

Lee, J., Y. Jung, J. H. Shin, H.K. Kim, B.C. Moon, D.H. Ryu and G.S. Hwang. 2014. Secondary metabolite profiling of *Curcuma* species grown at different locations using GC/TOF and UPLC/Q-TOF MS. Molecules. 19: 9535-9551.

Nirmal Babu, K., P.N. Ravindran and K.V. Peter.1997. Protocols for micropropagation of Spices and Aromatic Crops. Indian Institute of Spices Research, Calicut, Kerala, p.35.

Nirmal Babu, K., S.P. Geetha, D. Minoo, P. N. Ravindran and K.V. Peter.1999. *In vitro* conservation of germplasm. In. S P Ghosh (ed.) Biotechnology and its application in Horticulture. Narosa Publishing House, New Delhi, pp :106-129.

Nirmal Babu, K., S.P. Geetha, D. Minoo, P. N. Ravindran and K.V. Peter.1999. *In vitro* conservation of cardamom (*Elettaria cardamomum* Maton) germplasm. Plant genetic resources News Letter. 11:41-45.

Nishiyama, Y., T. Fukamizo, K. Yoneda and T. Araki. 2017. Complete amino acid sequence of a copper/zinc-superoxide dismutase from ginger rhizome. The Protein Journal. 36:98-107.

Nitsch, J. P. and C. Nitsch. 1969. Haploid plants from pollen grains. Science.163:85-87.

Niu, X., K. Lin, P.M. Hasegawa, R.A. Bressan and S.C. Weller. 1998. Transgenic peppermint (*Mentha× piperita* L.) plants obtained by cocultivation with *Agrobacterium tumefaciens*. Plant Cell Reports.17:165-171.

Nogoy, F. M., J.Y. Song, S. Ouk, S. Rahimi, S.W. Kwon, K.K. Kang and Y.G. Cho. 2016. Current applicable DNA markers for marker assisted breeding in abiotic and biotic stress tolerance in rice (*Oryza sativa* L.). Plant Breeding and Biotechnology. 4:271-284.

Obayemi, A. and S. Micheal. 2016. Roles of molecular markers in horticultural crop breeding programme. Global Journal of Animal Breeding and Genetics. 4:327-333.

Ohga,Y., M. Ono, and K.Furuno.1989. Somatic embryogenesis and plant regeneration from hypocotyls and cotyledons of Angelica acutiloba and Foeniculum vulgare. Report of the Kyushu Branch of the Crop Science Society of Japan. 56: 89-91.

Olszewska, D., I. Jedrzejczyk, P. Nowaczyk, S.Sendel and B. Gaczkowska. 2015. *In vitro* colchicine treatment of anther-derived pepper haploids. Bulgarian Journal of Agricultural Science. 21:806-810.

Owis, A. I., N.S. Abdelwahab and A.A. Abul-Soad. 2019. Analysis of phenolics in *Calligonum polygonoidesin vitro* cultured roots. Journal of Reports in Pharmaceutical Sciences.8:124-127.

Özkaynak, E., Z. Devran, E. Kahveci, S. Doğanlar, B. Başköylü, F.Doğan and M. Yüksel. 2014. Pyramiding multiple genes for resistance to PVY, TSWV and PMMoV in pepper using molecular markers. European Journal of Horticultural Science Science. 79 : 233–239.

Padmini K, M.N. Venugopal and S.J. Ankegowda. 2001. Heterosis for seedling characters in cardamom (*Elettaria cardamomum* Maton.). Journal of Spices and Aromatic Crops 9:145-151.

Pandey, S., M.K. Patel, A. Mishra and B. Jha. 2015. Physio-biochemical composition and untargeted metabolomics of cumin (*Cuminum cyminum* L.) make it promising functional food and help in mitigating salinity stress. PLoS One.10(12).Doi: 10.1371/journal.pone.0144469

Pandey, S., M.K. Patel, A. Mishra and B. Jha. 2016. In planta transformed cumin (*Cuminum cyminum* L.) plants, overexpressing the SbNHX1 gene showed enhanced salt endurance. PloS One.11(7).Doi: 10.1371/journal.pone.0159349

Pandit, A. A., R.A, Shah and A.M. Husaini. 2018.Transcriptomics: A time-efficient tool with wide applications in crop and animal biotechnology. Journal of Pharmacognosy and Phytochemistry.7:1701-1704.

Pank, F., Neumann, M. and H. Kruger,2000. Results of the selection of bitter fennel with small shaped fruits (*Foeniculum vulgare* Mill. ssp. vulgare var. vulgare) from cross progenies of different genotypes. Zeitschrift fur Arznei and Gewurzpflanzen. 5(1): 40-48 .

Parameswar, N.S. 1973. Floral biology of cardamom Mysore J. Agric. Sci. 7:205-213.

Parameswar, N.S. and R.Venugopal.1974. Study of flowering and anthesis in cardamom (*Elettaria cardamomum* Maton). Mysore J. Agric. Sci. 6:356-361.

Parida, R., S. Mohanty, and S. Nayak. 2018. Assessment of genetic diversity of *Curcuma aromatica* from eastern India using ISSR and RAPD markers. Journal of Applied and Advanced Research.3: 24-27.

Park, D. H., K.J. Sa, S.E. Lim, S.J. Ma and J.K. Lee. 2019. Genetic diversity and population structure of *Perilla frutescens* collected from Korea and China based on simple sequence repeats (SSRs). Genes & Genomics. 41: 1329-1340.

Partap, U., G. Sharma, M.B. Gurung, N., Chettri and E. Sharma. 2014. Large cardamom farming in changing climatic and socioeconomic conditions in the Sikkim Himalayas. ICIMOD Working Paper 2014/2. Kathmandu: ICIMOD.

Parthasarathy, V. A., B. Chempakam, T. J. Zachariah. 2008. Chemistry of Spices. Publisher: CABI, Wallingford, UK. pp.434. ISBN : 9781845934057

Parthasarathy, V.A. 2008. Spices research in India – a perspective. In: Solanki ZS, Parihar GN, Rathore BS, Kumhar SR, editors. Proceeding of National Workshop on Spices and Aromatic Plants. ARS, Mandor, pp. 1–20.

Parthasarathy, V.A., B. Chempakam, T.J. Zachariah. 2008. Chemistry of Spices. CABI, Wallingford, UK. pp.434.

Parthasarathy, V.A. 2008. Spices Research in India – a perspective. In: Solanki ZS, Parihar GN, Rathore BS, Kumhar SR, editors. Proceeding of National Workshop on Spices and Aromatic Plants. ARS, Mandor, pp. 1–20.

Parvathi, C and H.M. Chandrappa.1993. Preliminary screening of cardamom elite clones for thrips (Sciothripscardamomi Rank) damage in the field. J. Plantation crops. 2: 81-82.

Parvathy, V. A., V. P. Swetha, T. E.Sheeja and B. Sasikumar. 2015.Detection of plant-based adulterants in turmeric powder using DNA barcoding. PharmaceuticalBiology. 53:1774-1779.

Passricha, N., S. Saifi, S. Khatodia and N. Tuteja. 2016. Assessing zygosity in progeny of transgenic plants: current methods and perspectives. Journal of Biological Methods. 3(3). Doi: 10.14440/jbm.2016.114

Pasternak, T. P., E. Prinsen, F. Ayaydin, P. Miskolczi, G. Potters, H. Asard, and A. Fehér. 2002. The role of auxin, pH, and stress in the activation of embryogenic cell division in leaf protoplast-derived cells of alfalfa. Plant Physiology.129.1807-1819.

Patel, J.A., M.J.Patel, A.S. Bhanvadia, R.R.Acharya and M.K. Bhalala. 2001. Extent of natural cross pollination with gms lines in chilli (*Capsicum annuum* L.). Capsicum and Eggplant Newsl. 20: 35-37.

Patel, J.A., M.J.Patel, R.R.Acharya, A.S. Bhanvadia and M.K. Bhalala. 2004. Hybrid vigour, gene action and combining ability in chilli (*Capsicum annuum* L.) hybrids involving male sterile lines.The Indian Journal of Genetics and Plant Breeding. 64:81-82.

Patil, M.J., S.P. Ukey, B.N. Ghoderao. 2002. Performance of chilli against fruit rot in natural condition. Annals of Plant Physiology. 16; 202-203.

Patil,M.J., S.P.Ukey and B.N. Ghoderao. 2003. Screening of brinjal germplasm against phomopsis. Ann Pl Physiol.17:196-1977 .

Pattanshetti H.V. and A.B.N. Prasad.1972. Blossom biology, pollination and fruit set in cardamom Third Intl. Symp. subtropical and tropic, pp: 262-268.

Paul, R. and A. K . Datta. 2005. Gamma-rays and EMS induced macromutants in celery (*Apium graveolens* L.), fennel (*Foeniculum vulgare* Mill.) and ajowan (*Trachyospermum ammi* L.). Journal of Phytological Research. 18: 95-98.

Paun, O. and P. Schönswetter. 2012. Amplified fragment length polymorphism: an invaluable fingerprinting technique for genomic, transcriptomic, and epigenetic studies. p.75-87, In N.J.Sucher, J.R. Hennell and M.C. Carles, ed. Plant DNA Fingerprinting and Barcoding – Methods and Protocol, Humana Press, New York.

Pereira-Dias, L., S. Vilanova, A. Fita, J. Prohens and A. Rodríguez-Burruezo. 2019. Genetic diversity, population structure, and relationships in a collection of pepper (*Capsicum spp.*) landraces from the Spanish centre of diversity revealed by genotyping-by-sequencing (GBS). Horticulture Research. 6: 1-13.

Peter,K.V., P.N. Ravindran, K. Nirmal Babu, M.N. Venugopal, S.P. Geetha and D. Benny. 2001. Production of somaclones and somatic hybrids of cardamom for high yield and resistance to diseases. ICAR Project report. Indian Institute of Spices Research, Calicut, Kerala, India, pp. 120.

Philip, V.J., D. Joseph, G.S.Triggs and N.M.Dickinson.1992. Micropropagation of black pepper (*Piper nigrum* L.) through shoot tip cultures. Plant Cell Report. 12: 41-44.

Ponti, L. 2005. Transgenic crops and sustainable agriculture in the European context. Bulletin of Science, Technology and Society.25:289-305.

Portemer, V., C. Renne, A. Guillebaux and R. Mercier. 2015. Large genetic screens for gynogenesis and androgenesis haploid inducers in *Arabidopsis thaliana* failed to identify mutants. Frontiers in Plant Science.6:147.

Posch, A., B.M. van den Berg, W. Postel and A. Görg. 1992. Genetic variability of pepper (*Capsicum annuum* L.) seed proteins studied by 2-D electrophoresis with immobilized pH gradients. Electrophoresis.13:774-777.

Pradeep Kumar, T, J.L. Karihaloo, S. Archak and A. Baldev. 2003. Analysis of genetic diversity in *Piper nigrum* L. using RAPD markers. Genetic Res. Crop Evolution. 50: 469-475.

Pradeep Kumar, T., J.L. Karihaloo and S. Archak. 2001. Molecular characterization of *Piper nigrum* cultivars using RAPD markers. Current Sci. 8: 246-248.

Prakash, N., J.F. Brown and H. Yue.1994. An embryological study of Kava, *Piper methysticue*. Australian Journal of Botany. 42:231–237.

Prasad, B. C. N., K.M.Reddy and A.T.Sadashiva. 2003. Heterosis studies in chilli (*Capsicum annuum* L.). Indian Journal of Horticulture. 60: 69-74.

Prasath, D. and M.N. Venugopal. 2001. Genetic analysis for quantitative characters in cardamom. Indian J.Genet. 62:179-180.

Prasath, D. and M.N. Venugopal. and V.S. Korikanthimath.2001. Variability in cardamom. Indian J. Plant Genet. Resour. 14: 217-218.

Prasath, D., R.Karthika, N. T. Habeeba, E. J. Suraby, O. B. Rosana, A. Shaji .and M. Anandaraj. 2014. Comparison of the transcriptomes of ginger (*Zingiber officinale* Rosc.) and mango ginger (*Curcuma amada* Roxb.) in response to the bacterial wilt infection. Plos One. 9.Doi: 10.1371/journal.pone.0099731

Prohens, J. 2011. Plant breeding: a success story to be continued thanks to the advances in genomics. Frontiers in Plant Science. 2:51.

Purohit, S., K. Joshi, V. Rawat, I.D. Bhatt and S.K. Nandi.2019. Efficient plant regeneration through callus in *Zanthoxylum armatum* DC: an endangered medicinal plant of the Indian Himalayan region. Plant Biosystems-An International Journal Dealing with all Aspects of Plant Biology.154:1-7.

Qaderi, A., M. Omidi, A. Pour-Aboughadareh, P. Poczai, J. Shaghaghi, A. Mehrafarin, ….and A. Etminan. 2019. Molecular diversity and phytochemical variability in the Iranian poppy (*Papaver bracteatum* Lindl.): A baseline for conservation and utilization in future breeding programmes. Industrial Crops and Products.130: 237-247.

Qin, C., C. Yu, Y. Shen, X. Fang, L. Chen, J. Min and Y. Yang. 2014. Whole-genome sequencing of cultivated and wild peppers provides insights into Capsicum domestication and specialization. Proceedings of the National Academy of Sciences.111(14):5135-5140.

Qiu, G. F., L.W. Xiong, Z.K. Han, Z.Q. Liu, J.B. Feng, X.G. Wu, and L. Chen. 2017. A second generation SNP and SSR integrated linkage map and QTL mapping for the Chinese mitten crab *Eriocheir sinensis*. Scientific Reports.7(1): 1-11.

Quyen N.T, T.D. Hien, D.T. Oanh, N.Q. Ngoc and N.T. Nhung. 2019. Flower biology of black pepper (*Piper nigrum*) in Vietnam. Journal of Vietnam Agricultural Science and Technology. 1(4):19-23.

Raghavan, R., S. Elumalai, K.Nirmal Babu and S.Hittalmani. 2010. Molecular characterization of black pepper (*Piper nigrum*) using RAPD and SSR markers. Biosciences, Biotechnology Research Asia. 7: 1011-1015.

Raghunathachari, P., S.K. Nivas and L.D. Souza. 2012. *Agrobacterium* mediated transformation of a pure line variety of hot pepper, RCL 59M. International Journal of Science and Research. 3:2147-2150.

Raj, N.M., M. Murugan and R.C. Joseph. 2000. Evaluation of cardamom germplasm. Journal of spices and Aromatic Crops. 9:55-56.

Ram, H.H.1998.Vegetable Breeding: Principle and Practices. Kalyani Publishers, New Delhi, pp.135-36.

Ramadan, W.A., R.M. Shoaib, R.T. Ali and N.S. Abdel-Samea. 2019. Assessment of genetic diversity among some fennel cultivars (*Foeniculum vulgare* Mill.) by ISSR and SCoT Markers. African Journal of Biological Sciences.15: 219-234.

Ramadas, D. and L. Srinivas. 2011. Antioxidant effects of 28 KDA protein from turmeric (*Curcuma Longa* L). Asian Journal of Pharmaceutical and Clinical Research.4 (Suppl.1):75-79.

Rani, B. and V.K. Sharma. 2017. Transcriptome profiling: methods and applications-A review. Agricultural Reviews. 38(4):271-281.

Rani, D. and P.K. Dantu.2016. Sustained shoot multiplication and method for overcoming *in vitro* browning in medicinally important plant, *Piper chaba* hunt. Proceedings of the National Academy of Sciences, India Section B: Biological Sciences.86:407-413.

Rani, S.J. and R. Usha. 2013. Transgenic plants: Types, benefits, public concerns and future. Journal of Pharmacy Research.6:879-883.

Rao, N. K. 2004. Plant genetic resources: Advancing conservation and use through biotechnology. African Journal of Biotechnology.3:136-145.

Rao, S. R. and G.A. Ravishankar. 2002. Plant cell cultures: chemical factories of secondary metabolites. Biotechnology Advances. 20:101-153.

Rao, S.N.K., S. Narayanaswamy, E.K. Chacko and R. Doraiswamy.1982. Regeneration of plantlets from callus of cardamom. Proc. Indian Academy of Sci. (Plant Sci.). 91:37-41.

Ratnambal, M.J., P.N.Ravindran, M.K. Nair and K. Nirmal Babu.1990. Two high yielding selections of Karimunda. Spice India. 3:9-11.

Raveendar, S., K. J. Lee, M.J. Shin, G.T. Cho, J.R. Lee, K.H. Ma and J.W. Chung. 2017. Complete chloroplast genome sequencing and genetic relationship analysis of *Capsicum chinense* Jacq. Plant Breeding and Biotechnology.5:261-268.

Ravindran, P.N. 1991. Studies on black pepper and some of its wild relatives. Ph. D Thesis. University of Calicut.

Ravindran, P.N. 2000. Black Pepper. Harwood Academic Publishers, Singapore. p.540.

Ravindran, P.N. and K. Nirmal Babu. 1994. Genetic resources of black pepper. In KL Chadha and P Rethinam (eds.) Advances in Horticulture, Vol.9. Plantation Crops and Spices. Malhotra Publishing House, New Delhi, pp. 99-120.

Ravindran, P.N., K. Nirmal Babu and D. Joseph. 2002. Production of Haploids of Cardamom through Anther culture / Microspore culture. ICAR Project report. Indian Institute of Spices Research, Calicut, Kerala, India, pp. 51.

Ravindran, P.N., K. Nirmal Babu and K.N. Shiva. 2006. Black pepper. In: P.N. Ravindran, K. Nirmal Babu, K.N. Shiva and A.K. Johny (eds.) Advances in Spices Research, Agrobios, Jodhpur. pp. 215-291.

Ravindran, P.N., K. Nirmal Babu, B. Sasikumar, K.S.Krishnamoorthy. 2000. Botany and crop improvement of black pepper. In: Ravindran PN, editor. Black Pepper, Piper nigrum. Harwood Academic Publishers, Amsterdam, The Netherlands; pp. 23–142.

Ravindran, P.N., K. Nirmal Babu, K.V. Saji, S.P. Geetha, K. Praveen and G. Yamuna. 2004. Conservation of Spices genetic resources in invitro gene banks. ICAR Project report. Indian Institute of Spices Research, Calicut, Kerala, India, pp. 81.

Ravindran, P.N., K.V. Ramana, M.K.Nair, K.Nirmal Babu and C. Mohandas.1992. Pournami-a high yielding black pepper selection tolerant to root knot nematode (Meloidogyne incognita). J. Spices and Aromatic Crops. 1: 136-141.

Ravindran, P.N., M.J. Ratnambal and V.M. Muneer.1986. Differential radiation sensitivity of black pepper (*Piper nigrum* L.) cultivars. Planter, Kuala Lumpur, 62:515-518.

Ravindran, P.N., M.K. Nair and K. Nirmal Babu.1992. Panchami-A high yielding selection of black pepper. Spice India. 5: 11-13.

Ravindran, P.N., R. Balakrishnan, K.Nirmal Babu. 1997.Morphological studies on black pepper (*Piper nigrum* L.). I. Cluster analysis of black pepper cultivars. Journal of Spices and Aromatic Crops. 6:9–20.

Ravindran, P.N.and B. Sasikumar.1993. Variability in open pollinated progenies of black pepper (*Piper nigrum L.*). Journal Spices and Aromatic Crops. 2:60–65.

Ravishankar, G.A. and L.V. Venkataraman. 1993. Role of plant cell culture in food biotechnology: current trends, limitations and future prospects, p. 255-274, In: J.Prakash and R.L.M. Pierik, ed. Plant Biotechnology: Commercial Prospects and Problems. Oxford IBH, New Delhi.

Reddy, M. P., N. Sarla and E.A. Siddiq. 2002. Inter simple sequence repeat (ISSR) polymorphism and its application in plant breeding. Euphytica.128: 9-17.

Rema, J, C.Z. John and P.M. Mini. 1995. In vitro plant regeneration of economically important Piper species (*P. nigrum* L., *P. barberi* L., *P .longum* L., *P. chaba* Hunt). In: M.G. Nair (ed.), Proc. Seventh Kerala Science Congress, Palakkad, India. pp. 321-324.

Riahi, L., H. Chakroun, I. Klay, A.S. Masmoudi, A. Cherif and N. Zoghlami. 2018. Metabolomic fingerprint of *Mentha rotundifolia* L. Leaf tissues promotes this species as a potential candidate for sustainable production of biologically active molecules. Journal of Complementary and Integrative Medicine.16(2). Doi: 10.1515/jcim-2018-0048

Ribaut, J. M., M.C. De Vicente and X. Delannay. 2010. Molecular breeding in developing countries: challenges and perspectives. Current Opinion in Plant Biology. 13(2):213-218.

Robarts, D. W. and A.D. Wolfe. 2014. Sequence□related amplified polymorphism (SRAP) markers: A potential resource for studies in plant molecular biology[1]. Applications in Plant Sciences.2(7). Doi:10.3732/apps.1400017

Robinson, A. J., C.G. Love, J. Batley, G. Barker and D. Edwards. 2004. Simple sequence repeat marker loci discovery using SSR primer. Bioinformatics. 20(9):1475-1476.

Rout, G. R., S.K. Palai, S. Samantaray and P. Das. 2001. Effect of growth regulator and culture conditions on shoot multiplication and rhizome formation in ginger (*Zingiber officinale* Rosc.) *in vitro*. *In Vitro* Cellular & Developmental Biology-Plant.37:814-819.

Rumana A., K. L. W. Kumara, G. Senanayake, R.A.A.K Ranawaka, D.K.N.G. Pushpakumara and S. Geekiyanage. 2018. Flower morphological diversity of cinnamon (*Cinnamomum verum* Presl) in Matara District, Sri LankaOpen Agriculture. 3: 236–244.

Saccardo,F., F.Micozzi, C.Amici and M .Cardarelli.2002. Fennel hybrid F1 'grumolo' for improving the quality and resistance to stresses. Informatore Agrario. 58: 61-62.

Sahoo, A., B. Kar, S. Sahoo, A. Ray and S. Nayak. 2016. Transcriptome profiling of *Curcuma longa* L. cv. Suvarna. Genomics Data.10: 33-34.

Sahoo, A., S.Jena, , B. Kar, S.Sahoo, A Ray, S.Singh and S. Nayak. 2017. EST-SSR marker revealed effective over biochemical and morphological scepticism towards identification of specific turmeric (*Curcuma longa* L.) cultivars. 3 Biotech.7:84.

Sajina, A., D. Minoo, P. Geetha, K. Samsudeen, J. Rema, K. Nirmal Babu, P.N. Ravindran and K.V. Peter.1997. Production of synthetic seeds in few spice crops. In. Edison, S., Ramana, K.V., Sasikumar, B., Nirmal Babu, K. and Santhosh J. Eapen (Eds). Biotechnology of Spices, Medicinal and Aromatic Plants, Indian Society for Spices, Calicut, India, pp. 65-69.

Sakthipriya, M. and K.K. Sabu. 2018. Development and cross-genera transferability of ginger EST-SSR markers for cardamom. Current Bioinformatics.13:95-99.

Salvi, N. D., L. George and S. Eapen. 2002. Micropropagation and field evaluation of micropropagated plants of turmeric. Plant Cell, Tissue and Organ Culture (PCTOC). 68:143-151.

Sánchez-Bel, P., I. Egea, M.T. Sánchez-Ballesta, C. Martinez-Madrid, N. Fernandez-Garcia, F. Romojaro and F.B. Flores. 2012. Understanding the mechanisms of chilling injury in bell pepper fruits using the proteomic approach. Journal of Proteomics. 75:5463-5478.

Sasi, S. and A. I.Bhat. 2018. *In vitro* elimination of *Piper yellow mottle virus* from infected black pepper through somatic embryogenesis and meristem-tip culture. Crop Protection. 103: 39-45.

Sasi, S. and A.I. Bhat. 2016.Optimization of cyclic somatic embryogenesis and assessing genetic fidelity in six varieties of black pepper (*Piper nigrum* L). Journal of Medicinal Plants.4:109-115.

Sasikumar, B and K. Veluthambi.1994. Kanamycin sensitivity of cultured tissues of Piper nigrum L. J. Spices and Aromatic Crops. 3:158-162.

Sasikumar, B and K. Veluthambi.1996. Strain and vector specificity in Agrobacterium - black pepper interaction. J. Plantation Crops. 24 (Supplt): 597-602.

Sasikumar, B and K. Veluthambi.1996.Transformation of black pepper (Piper nigrum L.) using Agrobacterium Tiplasmid based vectors. Indian Perfumer. 40:13-16.

Sasikumar, B., Chempakam, B., George, J.K. et al. 1999. Characterization of two interspecific hybrids of Piper. Journal of Horticultural Science and Biotechnology. 74:125-31.

Sasikumar, B., K G. Johnson and P.N. Ravindran. 1992. Breeding behaviour of black pepper. Indian Journal of Genetics. 52:17–21.

Sasikumar, B., P.Haridas, J.K.George, K.V.Saji, T. John Zachariah, P.N. Ravindran et al. 2004. 'IISR Thevam', 'IISR Malabar Excel', 'IISR Girimunda' – three new black pepper clones. Journal of Spices and Aromatic Crops. 13:1–5.

Sastri, B.N. 1952. The wealth of India. Raw materials. D-E: 150-160.

Sato, H., S. Enomoto, S. Oka, K. Hosomi and Y. Ito. 1993. Plant regeneration from protoplasts of peppermint (*Mentha piperita* L.). Plant Cell Reports.12:546-550.

Savelkoul, P. H. M., H.J.M. Aarts, J. De Haas, L. Dijkshoorn, B. Duim, M. Otsen and J.A. Lenstra. 1999. Amplified-fragment length polymorphism analysis: the state of an art. Journal of Clinical Microbiology. 37(10): 3083-3091.

Schiavone, F. M. and T.J. Cooke. 1987. Unusual patterns of somatic embryogenesis in the domesticated carrot: developmental effects of exogenous auxins and auxin transport inhibitors. Cell Differentiation. 21(1):53-62.

Schmidt, E. D., F. Guzzo, M.A. Toonen and S.C. De Vries. 1997. A leucine-rich repeat containing receptor-like kinase marks somatic plant cells competent to form embryos. Development.124:2049-2062.

Schneider, M. V. and S. Orchard. 2011. Omics technologies, data and bioinformatics principles. p. 3-30, In B. Meyer, ed. Bioinformatics for Omics Data. Humana Press.

Sebastin, R., K.J. Lee, G.T. Cho, M.J. Shin, S.H. Kim, D.Y. Hyun and J.R. 2019. The complete chloroplast genome sequence of a Bolivian wild chili pepper, *Capsicum eximium* Hunz. (Solanaceae). Mitochondrial DNA Part B. 4:1634-1635.

Selvakumar, T., T.Thangaselvabai, K.R. Sudha and R.Balakumbahan. 2011. KVK Tamil Nadu Agriculture University Pechiperai, Agri. Review. 32:283-293,.

Semagn, K., Å. Bjørnstad and M.N. Ndjiondjop. 2006a. An overview of molecular marker methods for plants. African Journal of Biotechnology.5:2540-2568.

Semagn, K., Bjørnstad, Å.,& Ndjiondjop, M. N. 2006b. Progress and prospects of marker assisted backcrossing as a tool in crop breeding programs. African Journal of Biotechnology. 5:2588-2603.

Senan, S., D. Kizhakayil, T.E. Sheeja, B. Sasikumar, A.I. Bhat and V.A. Parthasarathy. 2013. Novel polymorphic microsatellite markers from turmeric, *Curcuma longa* L.(Zingiberaceae). Acta Botanica Croatica.72: 407-412.

Sengar, R. S., R. Chaudhary and S.K. Tyagi. 2010. Present status and scope of floriculture developed through different biological tools. Research Journal of Agricultural Science.1:306-314.

Senizza, B., G. Rocchetti, S. Ghisoni, M. Busconi, M.D.L.M. Pascual, J.A. Fernandez, ...and M. Trevisan. 2019. Identification of phenolic markers for saffron authenticity and origin: An untargeted metabolomics approach. Food Research International.126. Doi: 10.1016/j. foodres.2019.108584

Servin, B., O.C. Martin and M. Mézard. 2004. Toward a theory of marker-assisted gene pyramiding. Genetics, 168(1):513-523.

Sethi, S. and E. Brietzke. 2016. Omics-based biomarkers: application of metabolomics in neuropsychiatric disorders. International Journal of Neuropsychopharmacology.19:1-13.

Sevindik, B. and Y.Y. Mendi. 2016. Somatic Embryogenesis in *Crocus sativus* L. p. 351-357, In M. A. Germanà and M. Lambardi, ed. *In Vitro* Embryogenesis in Higher Plants, Humana Press, New York, NY.

Shajahan, A., C.S. Raju, C. Thilip, K. Varutharaju, K. Faizal, V.M. Mehaboob and A. Aslam. 2016. Direct and indirect somatic embryogenesis in mango ginger (*Curcuma amada* Roxb.). p.367-379, In V. M. Loyola-Vargas and N. Ochoa-Alejo, ed. Somatic Embryogenesis: Fundamental aspects and applications. Springer, Cham.

Shaji, P., M. Anandaraj and Y. R. Sharma. 1998. Comparative study of protoplast isolation and development in *Piper nigrum* and*P. colubrinum*. In NM Mathew and CK Jacob (Eds.) Developments in plantation crops research, Allied Publishers, New Delhi. pp. 51-53.

Shalini, S., A. Singla, M. Goyal, V.Kaur and P. Kumar. 2018.p.344-360, Crop Improvement for Sustainability, In P.K.Yadav, S.Kumar, S.Kumar and R.C.Yadav, eds. Omics in Agriculture: Applications, Challenges and Future Perspectives. Daya Publishing House.

Shamina, A., T.J. Zachariah, B. Sasikumar and J.K. George. 1998. Biochemical variation in turmeric (*Curcuma longa* Linn.) accessions based on isozyme polymorphism. The Journal of Horticultural Science and Biotechnology.73:479-483.

Shankarnag, B., M.B.Madalageri, M. Ravindra. 2006. Manifestation of heterosis for growth, earliness and early green fruit yield in chilli.Indian J Horticulture. 63 : 410-414.

Shankarnag, B., M.B.Madalageri, S.C.Hiremath, M.P.Patil, M.C. Wali. 2005. Heterosis for fruit and yield parameters in chilli (*Capsicum annuum* L.). Karnataka Journal of Horticulture. 1: 7-11.

Sharangi, A.B., K. Alam, R. Chatterjee and A.Pariari. 2003. Evaluation of fennel (*Foeniculum vulgare* M.) germplasm for growth and yield. Environment and Ecology. 21: 477-479 .

Sharma, N. and S.L. Harikumar . 2013. Use of genomics and proteomics in pharmaceutical drug discovery and development: a review. International Journal of Pharmacy and Pharmaceutical Science. 5: 24-28.

Sharma, A.K. and N.K. Bhattacharyya.1959. Cytology of several members of Zingiberaceae and the study of the inconsistency of their chromosome complements. La.Cellule 59:297-346.

Sharma, G., U. Partap, D.P.Sharma. 2019. Pollination biology of large cardamom (*Amomum subulatum Roxb.*) with special emphasis on honey bees (*Apisspp.*) and bumble bees (*Bombus spp.*) pollinators. Tropical Ecology. 60:507–517.

Sharma, M., A. Ahuja, R. Gupta and S. Mallubhotla. 2015. Enhanced bacoside production in shoot cultures of *Bacopa monnieri* under the influence of abiotic elicitors. Natural Product Research.29(8):745-749.

Sharma, M., A. Sharma, A. Kumar and S.K. Basu. 2011. Enhancement of secondary metabolites in cultured plant cells through stress stimulus. American Journal of Plant Physiology. 6:50-71.

Sharma, M., M. Sharma, R.K. Salgotra, M. Gupta, A.K. Singh and L.M. Gupta.2019. Development of an effective protocol for *in vitro* multiplication of peppermint (*Mentha piperita*). Indian Journal of Agricultural Sciences.89:223-226.

Sharma, T. R. and B.M.Singh.1997.High-frequency *in vitro* multiplication of disease-free *Zingiber officinale* Rosc. Plant Cell Reports.17:68-72.

Sharma,M.V. and J.E. Armstrong. 2013. Pollination of Myristica and other nutmegs in natural populations. Tropical Conservation Science.6(5):593-607.

Sharmin, A., M.E. Hoque, M.M. Haque and F. Khatun. 2018. Molecular diversity analysis of some chilli (*Capsicum spp.*) genotypes using SSR markers. American Journal of Plant Sciences.9:368-379.

Sheeja, T. E., C. Sabeesh, O.V. Shabna, R.S. Shalini and B. Krishnamoorthy.2013. Genetic diversity analysis of Myristica and related genera using RAPD and ISSR markers. Journal of Spices and Aromatic Crops.22:38-46.

Shiraki, T., S. Kondo, S. Katayama, K. Waki, T. Kasukawa, H. Kawaji and S. Fukuda. 2003. Cap analysis gene expression for high-throughput analysis of transcriptional starting point and identification of promoter usage. Proceedings of the National Academy of Sciences.100: 15776-15781.

Shirgurkar, M. V., V. B. Naik, S. von Arnold, R.S. Nadgauda and D. Clapham. 2006. An efficient protocol for genetic transformation and shoot regeneration of turmeric (*Curcuma longa* L.) via particle bombardment. Plant Cell Reports. 25: 112-116.

Shivakumara, T. N., R. Sreevathsa, P.K. Dash, M.S. Sheshshayee, P.K. Papolu, U. Rao, and M. UdayKumar, M. 2017. Overexpression of Pea DNA Helicase 45 (PDH45) imparts tolerance to multiple abiotic stresses in chili (*Capsicum annuum* L.). Scientific Reports.7:2760. Doi:10.1038/s41598-017-02589-0

Shou, H., B.R. Frame, S.A. Whitham and K. Wang. 2004. Assessment of transgenic maize events produced by particle bombardment or *Agrobacterium*-mediated transformation. Molecular Breeding. 13(2): 201-208.

Shrestha, S. L., B.P. Luitel and W.H. Kang. 2011. Agro-morphological characterization of anther derived plants in sweet pepper (*Capsicum annuum* L. cv. Boogie). Horticulture, Environment, and Biotechnology.52(2):196-203.

Shrikhande, M., S.R. Thengane and A.F. Mascarenhas.1993.Somatic embryogenesis and plant regeneration in *Azadirachta indica* A. Juss. *In Vitro* Cellular & Developmental Biology-Plant.29:38-42.

Shylaja, M.R., S.G. Nair, P.A.Nazeem, V.K.Mallika and M.K. Mathew.1994. In vitro screening of black pepper for tolerance to *Phytophthora capsici*. PLACROSYM XI, Calicut: Nov - 3 Dec 1994. pp.24-30.

Sihanat, A., K. Rungsihirunrat and O. Theanphong. 2016. Phylogenetic relationships of *Kaempferia* species based on AFLP marker. Thai Journal Botany. 8:141-155.

Sim, S.L. 1993. Clonal selection and hybridization in pepper. In: Ibrahim MY, Bong CF, Ipore IB, editors. The Pepper Industry: Problems and Prospects. UniversitiPertanim Malaysia Bintulu Campus, Bintulu, Sarawak, Malaysia. pp. 48–57.

Sim, S.L., Jafar, R., Grierson, D., Power, J.B. and Davery, M.R.1995. Application of molecular biology and genetic manipulation to pepper (*Piper nigrum* L.) breeding: present state and prospects. In: Abstracts FAO/IAEA International symposium on the use of Induces Mutations and Molecular techniques for crop improvement, Vienna, Austria.

Sim, S.L., R. Jafar, J.B. Power and M.R. Davey.1998. Development of an Agrobacterium-mediated transformation system for black pepper (*Piper nigrum* L.). Acta Horticulturae. 461: 349-354.

Sim, S.L., T.H. Wong, T.K. Kueh, A.D. Paulus. 1993.Comparative performance of three varieties of pepper. In: Ibrahim MY, Bong CT, Ipor IB, editors. The Piper Industry: Problems and Prospects. University Pertanian Malaysia, Malaysia. pp. 2–14.

Šimić, D., T. Ledenčan, A. Jambrović, Z. Zdunić, J. Brkić, A. Brkić, and I. Brkić. 2009. SNP and SSR marker analysis and mapping of a maize population. Genetika.41: 237-246.

Singh, J. 1993. Improvement in chillies. In: Advance in Horticulture, ed. By Chadha, k.L. and Kallo, G., Vol 5, MBH, New Delhi, pp. 69-87.

Singh, J., A.K.Singh and S.P.S. Khanuja. 2003. Medicinal plants: India's opportunities. Pharma Bioworld.1: 59-66.

Singh, S., A. Kuanar, S. Mohanty, E. Subudhi and S. Nayak. 2011. Evaluation of phytomedicinal yield potential and molecular profiling of micropropagated and conventionally grown turmeric (*Curcuma longa* L.). Plant Cell, Tissue and Organ Culture (PCTOC).104:263-269.

Singhal, N., M. Kumar, P.K. Kanaujia and J.S. Virdi. 2015. MALDI-TOF mass spectrometry: an emerging technology for microbial identification and diagnosis. Frontiers in Microbiology. 6: 791.

Sinoj, J., T.E. Sheeja, R.S. Bhai, E.J. Suraby, D. Minoo, J. Soji, ... and K.N. Babu. 2014. Somatic embryogenesis and transgenic development in black pepper for delayed infection and decreased spread of foot rot caused by *Phytophthora capsici*. Journal of Plantation Crops. 42: 20-28.

Sinu P.A. and K.R.Shivanna 2007. Pollination biology of large cardamom (*Amomum subulatum*). Current Science. 93(4): 548-552.

Sivashankari, S. and P. Shanmughavel. 2007. Comparative genomics-a perspective. Bioinformation. 1:376-378.

Skolnick, J., J.S. Fetrow and A. Kolinski. 2000. Structural genomics and its importance for gene function analysis. Nature Biotechnology.18:283-287.

Skoog, F. and C. Miller. 1957. Chemical regulation of growth and organ formation in plant tissues cultured. *In Vitro* Symp Soc Exp Biol.11:118-131.

Slater, A. T., N.O. Cogan and J.W. Forster. 2013. Cost analysis of the application of marker-assisted selection in potato breeding. Molecular Breeding.32:299-310.

Ślusarkiewicz-Jarzina, A., H. Pudelska, J. Woźna and T. Pniewski. 2017. Improved production of doubled haploids of winter and spring triticale hybrids via combination of colchicine treatments on anthers and regenerated plants. Journal of Applied Genetics.58:287-295.

Smith, J. S. C. and O.S.Smith. 1992. Fingerprinting crop varieties.p.85-140, In D.L. Sparks, ed. Advances in Agronomy (Vol. 47), Academic Press.Elsevier.

Smýkalová, I., P. Šmirous, M. Kubošiová, N. Gasmanová and M. Griga. 2009. Doubled haploid production via anther culture in annual, winter type of caraway (*Carum carvi* L.). Acta Physiologiae plantarum.31:21-31.

Solomon, A. M., K. Han, J.H. Lee, H.Y. Lee, S. Jang and B.C. Kang. 2019. Genetic diversity and population structure of Ethiopian *Capsicum* germplasms. PloS One.14(5).Doi.10.1371/journal.pone.0216886

Song, W. S., S.D. Oh, H.M. Cho.1991. Adventive embryogenesis and plant regeneration from stem and leaf segment of Foeniculum vulgare Gaertner. 3. Effect of plant growth regulators on plant regeneration from adventive embryos. Research Reports of the Rural Development Administration, Biotechnology. 33: 60-65 .

Southgate, E. M., M.R. Davey, J.B. Power and R. Marchant. 1995. Factors affecting the genetic engineering of plants by microprojectile bombardment. Biotechnology Advances. 13:631-651.

Sreedevi, M., S. Syamkumar and B. Sasikumar. 2005. Molecular and morphological characterization of new promising black pepper (*Piper nigrum* L.) lines. J. Spices Aromatic Crops. 14:1-9.

Sreekumar, B. 2008. Global pepper economy-trends and forecast. In: Krishnamurthy KS, Kandiannan K, Suseela Bhai R, Saji KV,Parthasarathy VA, editors. National Seminar on Piperaceae –Harnessing Agr-technologies for Accelerated Production of Economically Important Spices. Calicut, pp. 176–88.

Srivastava, A., S. Gupta, K. Shanker, N. Gupta, A.K. Gupta and R.K. Lal. 2020. Genetic diversity in Indian poppy (*P. somniferum* L.) germplasm using multivariate and SCoT marker analyses. Industrial Crops and Products.144. Doi: 10.1016/j.indcrop.2019.112050

Stephin, S., A. Gangaprasad, S.P.Mathew, S. Muthukrishnan, S. 2020. Enhanced *in vitro* shoot multiplication of *Piper sarmentosum* by suppression of apical dominance. Proceedings of the National Academy of Sciences, India Section B: Biological Sciences. 90(1):87-94.

Subramanyam, K., K.V. Sailaja, K. Subramanyam, D.M. Rao and K. Lakshmidevi. 2011. Ectopic expression of an osmotin gene leads to enhanced salt tolerance in transgenic chilli pepper (*Capsicum annum* L.). Plant Cell, Tissue and Organ Culture (PCTOC). 105:181-192.

Sukumara Pillai, V., K.K. Ibrahim, S.Sasikumaran. 1994. Advances in horticulture. P.293-296, In: Chadha K.L. and P. Rethinam eds. Plantation Crops and Spices, Part I. Volume 9. Malhotra Publishing House, New Delhi.

Suma, B., R. Keshavachandran and E.V. Nybe. 2008. *Agrobacterium tumefaciens* mediated transformation and regeneration of ginger (*Zingiber officinale* Rosc.). Journal of Tropical Agriculture. 46:38-44.

Supena, E. D. J., S. Suharsono, E. Jacobsen and J.B.M. Custers, 2006. Successful development of a shed-microspore culture protocol for doubled haploid production in Indonesian hot pepper (*Capsicum annuum* L.). Plant Cell Reports. 25:1-10.

Sütyemez, M., G. Güvenç, Ş. B., Bükücü and A. Özcan . 2019. The Determination of Genetic Diversity among some Sumac (*Rhus coriaria* L.) Genotypes. Erwerbs Obstbau.61: 355-361.

Suzuki, M., T. Akihama and M. Ishikawa. 2005. Cryopreservation of encapsulated gentian axillary buds following 2 step-preculture with sucrose and desiccation. Plant Cell, Tissue and Organ Culture (PCTOC). 83:115-121.

Sze, S. C. W., J.X. Song, R.N. S. Wong, Y.B. Feng, T.B. Ng, Y. Tong and K.Y.B Zhang. 2008. Application of SCAR (sequence characterized amplified region) analysis to authenticate *Lycium barbarum* (wolfberry) and its adulterants. Biotechnology and Applied Biochemistry. 51: 15-21.

Taira, S., S. Shimma, I. Osaka, D. Kaneko, Y. Ichiyanagi, R. Ikeda,.....and K. Komatsu. 2012. Mass spectrometry imaging of the capsaicin localization in the capsicum fruits. International Journal of Biotechnology. 1: 61-65.

Takebe, I., G. Labib and G.Melchers.1971.Regeneration of whole plants from isolated mesophyll protoplasts of tobacco. Naturwissenschaften.58:318-320.

Tan, S. Z., P. Begley, G. Mullard, K.A. Hollywood and P.N. Bishop. 2016. Introduction to metabolomics and its applications in ophthalmology. Eye.30:773-783.

Tanweer, S., T. Mehmood, S. Zainab, Z. Ahmad and A. Shehzad. 2020. Comparison and HPLC quantification of antioxidant profiling of ginger rhizome, leaves and flower extracts. Clinical Phytoscience.6:12. Doi: 10.1186/s40816-020-00158-z

Taranto, F., N. D'Agostino, B. Greco, T. Cardi and P. Tripodi. 2016. Genome-wide SNP discovery and population structure analysis in pepper (*Capsicum annuum*) using genotyping by sequencing. BMC Genomics.17: 943. Doi: 10.1186/s12864-016-3297-7

Tautz, D. 1989. Hypervariability of simple sequences as a general source for polymorphic DNA markers. Nucleic Acids Research.17: 6463-6471.

Teshome, I., S. Teshome, T. Soromessa and T. Feyissa. 2016. Development of an efficient *in vitro* propagation protocol for *Satureja punctata*-A rare aromatic and medicinal plant. Taiwania.61:41–48.

Thabuis, A., A. Palloix, B. Servin, A.M. Daubeze, P. Signoret and V. Lefebvre. 2004. Marker-assisted introgression of 4 *Phytophthora capsici* resistance QTL alleles into a bell pepper line: validation of additive and epistatic effects. Molecular Breeding.14: 9-20.

Thangaselvabai, T., R.R.Kennedy, J.J. Prem, and M. Jayaseka. 2010. Clove (*Syzigium aromaticum*) – The spicy flower bud of significance- A Review. Agricultural Reviews. 31(1):40-47

Theiler Hedtrich, R. and A.C. Kagi. 1992. Cloning *in vitro* and somatic embryogenesis in *Foeniculum vulgare* Mill. (fennel) of 'Zefa Fino' and 'Zefa Tardo'. Acta Horticulturae. (300): 287-291.

Thorpe, T. A. 2007. History of plant tissue culture. Molecular Biotechnology. 37:169-180.

Tomar, U.K. and Dantu, P.K. 2010. Protoplast culture and somatic hybridization.p.876-891, In G.Tripathi, ed. Cellular and Biochemical Science. I.K. International House Pvt. Ltd., New Delhi.

Toonen, M.A.J. and S.C. De Vries. 1996. Initiation of somatic embryos from single cells. In: T.L.Wang and A.Cuming, ed. Embryogenesis: the Generation of a Plant. Oxford, Bios Scientific Publishers.pp. 173-189.

Torelli, A., M. Marieschi and R. Bruni. 2014. Authentication of saffron (*Crocus sativus* L.) in different processed, retail products by means of SCAR markers. Food Control. 36:126-131.

Torres, A. M., C.M. Avila, N. Gutierrez, C. Palomino, M.T. Moreno and J.I. Cubero. 2010. Marker-assisted selection in faba bean (*Vicia faba* L.). Field Crops Research.115:243-252.

Torres, K. C. 1989. Stages of micropropagation. p.52-65, In K. C. Torres, ed. Tissue Culture Techniques for Horticultural Crops. Springer, Boston, MA.

Torricelli, R., J. Yousefi, E. Albertini, R. Venanzoni and Y.G. Hosseinzadeh. 2019. Morphological and molecular characterization of Italian, Iranian and Spanish saffron (*Crocus sativus* l.) Accessions. Applied Ecology and Environmental Research.17:1875-1887.

Trindade, H. 2010. Molecular biology of aromatic plants and spices.A review. Flavour and Fragrance Journal.25:272-281.

Truong, H. T. H., J.H. Kim, M.C. Cho, S.Y. Chae and H.E. Lee. 2013. Identification and development of molecular markers linked to Phytophthora root rot resistance in pepper (*Capsicum annuum* L.). European Journal of Plant Pathology.135: 289-297.

Tulsani, N. J., R. Hamid, F. Jacob, N.G. Umretiya, A. K. Nandha, R.S. Tomar, R. S. and B. A. Golakiya. 2020. Transcriptome landscaping for gene mining and SSR marker development in Coriander (*Coriandrum sativum* L.). Genomics. 112:1545-1553.

Tyagi, R. K., R. Goswami, R. Sanayaima, R. Singh, R. Tandon and A. Agrawal. 2009. Micropropagation and slow growth conservation of cardamom (*Elettaria cardamomum* Maton). *In Vitro* Cellular & Developmental Biology-Plant. 45:721-729.

Umadevi, P., K. V.Saji, E. J. Suraby. 2015. Meristem culture for rapid regeneration in Black pepper (*Piper nigrum* Linn.) Annals of Plant Sciences. 4:1029-1032.

Ünlü, M., M.E. Morgan and J.S. Minden .1997.Difference gel electrophoresis. A single gel method for detecting changes in protein extracts. Electrophoresis.18:2071-2077.

Usman, M. G., M.Y. Rafii, M.Y. Martini, O.A. Yusuff, M.R. Ismail and G. Miah. 2018. Introgression of heat shock protein (Hsp70 and sHsp) genes into the Malaysian elite chilli variety Kulai (*Capsicum annuum* L.) through the application of marker-assisted backcrossing (MAB). Cell Stress and Chaperones.23:223-234.

Usman, M. G., M.Y. Rafii, O. Yusuff, M.Y. Martini, M.R. Ismail and R. Ridzuan. 2020. Molecular confirmation of candidate Hsp70 gene associated with heat tolerance in BC3F2 advanced backcross lines and their phenotypic resemblance with recurrent chilli Kulai. Acta Agriculturae Scandinavica, Section B—Soil & Plant Science.70:252-264.

Vallejos, C. E.1983. Enzyme activity staining.p 469-516, In S. D. Tanksley and T. J. Orton, ed. Developments in Plant Genetics and Breeding (Vol. 1, Part A).Elsevier.

Vanaja, T., Neema, V.P., Mammootty, K.P. and R. Rajeshkumar. 2008. Development of a promising interspecific hybrid in black pepper *(Piper nigrum)* for Phytopthora foot rot resistance. Euphytica,161:437-45.

Vandekerckhove, M., B. Van Droogenbroeck, M. De Loose, I. Taverniers, E. Daeseleire, P. Gevaert, and C. Van Poucke. 2017. Development of an LC-MS/MS method for the detection of traces of peanut allergens in chili pepper. Analytical and Bioanalytical Chemistry. 409:5201-5207.

Vanisree, M., C.Y.Lee, S.F. Lo, S.M. Nalawade, C.Y. Lin and H.S. Tsay. 2004. Studies on the production of some important secondary metabolites from medicinal plants by plant tissue cultures. Botanical Bulletin of Academia Sinica.45: 1-22.

Varghese, J. M. and A.I. Bhat. 2011. An efficient *Agrobacterium*-mediated transformation protocol for black pepper (*Piper nigrum* L.) using embryogenic mass as explant. Journal of Crop Science and Biotechnology.14:247-254.

Varghese, S., G.E. Thomas and G.Thomas. 2018. AFLP analysis reveals exceptionally narrow genetic background in ginger (*Zingiber officinale* Rosc.). International Journal of Engineering, Science and Mathematics.7:19-24.

Varshney, R. K. and A. Dubey. 2009.Novel genomic tools and modern genetic and breeding approaches for crop improvement. Journal of Plant Biochemistry and Biotechnology.18:127-138.

Varshney, R. K., Hoisington, D. A., Nayak, S. N., and A. Graner. 2009. Molecular plant breeding: methodology and achievements. p.283-304, In J.P. Gustafson, P. Langridge and D.J. Somers, eds. Plant Genomics, Humana Press.

Vasil, V., and I.K. Vasil. 1980. Isolation and culture of cereal protoplasts. II. Embryogenesis and plantlet formation from protoplasts of *Pennisetum americanum*. Theor. Appl. Genet. 56: 97-99.

Vasile, L., Z. Maria, V. Simona and A. Eliza. 2011. Use of nodal explants in " *in vitro*" micropropagation of *Mentha piperita* L. Fascicula Protecţia Mediului.16:247-251.

Vatsya B, Dinesh K, Kundapurkar AR and Bhaskaran S, 1987. Large scale plant formation of cardamom (Elettariacardamomum) by shoot bud cultures. Plant Physiology and Biochemistry 14:14 -19.

Venkataiah, P., P. Bhanuprakash, S.S. Kalyan and K. Subhash. 2016.Somatic embryogenesis and plant regeneration of *Capsicum baccatum* L. Journal of Genetic Engineering and Biotechnology.14:55-60.

Venugopal, M.N. 1999. Natural disease escapes as source of resistance against cardamom mosaic virus causing katte disease of cardamom (Elettaria cardamomum Maton). Journal of Spices and Aromatic Crops. 8:145-151.

Vijayan, K. 2005. Inter simple sequence repeat (ISSR) polymorphism and its application in mulberry genome analysis. International Journal of Industrial Entomology. 10 : 79-86.

Villa, C., J. Costa, L. Meira, M.B.P. Oliveira and I. Mafra. 2016. Exploiting DNA mini-barcodes as molecular markers to authenticate saffron (*Crocus sativus* L.). Food Control.65:21-31.

Visarada, K. B. R. S., K. Meena, C. Aruna, S. Srujana, N. Saikishore and N. Seetharama. 2009. Transgenic breeding: perspectives and prospects. Crop Science.49(5):1555-1563.

Vivek, P. J., N. Tuteja and E.V. Soniya. 2013. CDPK1 from ginger promotes salinity and drought stress tolerance without yield penalty by improving growth and photosynthesis in *Nicotiana tabacum*. PLoS One.8(10). Doi:10.1371/journal.pone.0076392

Vu, H. T. T., D.D. Le, A.M. Ismail and H.H. Le. 2012. Marker-assisted backcrossing (MABC) for improved salinity tolerance in rice ('*Oryza sativa*' L.) to cope with climate change in Vietnam. Australian Journal of Crop Science.6:1649-1654.

Wahyuno, D.; Manohara, D.; Ningsih,S.D.; Setijono,R.T. (2010). Development of improved black pepper variety resistant to foot rot disease caused by *Phytophthora capsici*. Journal Peneliti and an Pengembangan Pertanian. 29(3): 86-95.

Wang, J., J. Lv, Z. Liu, Y. Liu, J. Song, Y. Ma,.... and N. Juntawong. 2019. Integration of Transcriptomics and Metabolomics for Pepper (*Capsicum annuum* L.) in Response to Heat Stress. International Journal of Molecular Sciences. 20: 5042.

Wang, Z., M. Gerstein and M. Snyder. 2009. RNA-Seq: a revolutionary tool for transcriptomics. Nature Reviews Genetics.10: 57-63.

Wankhade, R. R. and M.S. Wadikar. 2018. Identification of sequence-characterized amplified regions (SCARs) markers linking resistance to powdery mildew in chilli pepper (*Capsicum annuum* L.). African Journal of Agricultural Research.13:2771-2779.

Wannapinpong, S., K. Srikulnath, A. Thongpan, K. Choowongkomon and S. Peyachoknagul.2015. Molecular cloning and characterization of the CHS gene family in turmeric (*Curcuma longa* Linn.). Journal of Plant Biochemistry and Biotechnology.24:25-33.

Weeks, A. R., T. Van Opijnen and J.A.J. Breeuwer. 2000. AFLP fingerprinting for assessing intraspecific variation and genome mapping in mites. Experimental & Applied Acarology. 24:775-793.

Weng, C., T.L. Kubisiak and M. Stine.1998. SCAR markers in a longleaf pine x slash pine F 1 family. Forest Genetics. 5 : 239-247.

Werner, K., W. Friedt and F. Ordon. 2005. Strategies for pyramiding resistance genes against the barley yellow mosaic virus complex (BaMMV, BaYMV, BaYMV-2). Molecular Breeding.16: 45-55.

Wierdl, M., M. Dominska and T.D. Petes. 1997. Microsatellite instability in yeast: dependence on the length of the microsatellite. Genetics.146:769-779.

Wijerathna, Y. M. A. M. 2015. Marker assisted selection: biotechnology tool for rice molecular breeding. Advances in Crop Science and Technology.3:187. doi:10.4172/2329-8863.1000187

Windsor, A. M., B.M. Ott, N. Zhang, J. Wen, E. Hsu and S.M. Handy. 2019. Full Chloroplast Genome Sequence of the Economically Important Dietary Supplement and Spice *Curcuma longa*. Microbiology Resource Announcements.8: Doi:10.1128/MRA.00576-19.

Winter, P. and G. Kahl. 1995. Molecular marker technologies for plant improvement. World Journal of Microbiology and Biotechnology.11: 438-448.

Wolf, J. B. 2013. Principles of transcriptome analysis and gene expression quantification: an RNA seq tutorial. Molecular Ecology Resources.13:559-572.

Wu, Z., J. Cheng, C. Qin, Z. Hu, C. Yin and K. Hu. 2013.Differential proteomic analysis of anthers between cytoplasmic male sterile and maintainer lines in *Capsicum annuum* L. International Journal of Molecular Sciences. 14:22982-22996.

Xiao, Y., G. Niu and T. Kozai. 2011. Development and application of photoautotrophic micropropagation plant system. Plant Cell, Tissue and Organ Culture (PCTOC).105:149-158.

Xie, L. B., X. Wang, M. Peng, Y.G. Zhou, L.X. Chen, L.X.Liu and Y.H. Guo. 2017. Comparative proteome analysis in hot pepper (*Capsicum annuum* L.) after space flight. Phyton, International Journal of Experimental Botany. 86:236-245.

Xiong, F., R. Zhong, Z. Han, J. Jiang, L. He, W. Zhuang and R. Tang, R. 2011. Start codon targeted polymorphism for evaluation of functional genetic variation and relationships in cultivated peanut (*Arachis hypogaea* L.) genotypes. Molecular Biology Reports.38: 3487-3494.

Xu ,Y .2010. Molecular plant breeding. CAB International, Wallingford, UK/Cambridge, MA.

Xu, S., X. Ge, S. Li, X. Guo, D. Dai and T. Yang. 2019. Discrimination of Different Parts of Saffron by Metabolomic Based Ultra Performance Liquid Chromatography Coupled with High Definition Mass Spectrometry. Chemistry & Biodiversity.16(10). Doi:10.1002/cbdv.201900363

Xu, X., M. Peng, Z. Fang and X. Xu. 2000. The direction of microsatellite mutations is dependent upon allele length. Nature Genetics. 24(4): 396-399.

Yamuna, G., V. Sumathi, S.P. Geetha, K. Praveen, N. Swapna and K. Nirmal Babu. 2007. Cryopreservation of *in vitro* grown shoots of ginger (*Zingiber officinale* Rosc.). Cryo Letters. 28(4):241-252.

Yan, G., H. Liu, H. Wang, Z. Lu, Y. Wang, D. Mullan and C. Liu.2017. Accelerated generation of selfed pure line plants for gene identification and crop breeding. Frontiers in Plant Science.8:1786.Doi:10.3389/fpls.2017.01786

Yang, L., S. Fu, M.A. Khan, W. Zeng and J. Fu .2013.Molecular cloning and development of RAPD-SCAR markers for *Dimocarpus longan* variety authentication. Springer Plus. 2: 501.

Yang, X. and X. Zhang. 2010. Regulation of somatic embryogenesis in higher plants. Critical Reviews in Plant Science.29:36-57.

Yang, X., K. Gao, Z. Chen, X. Yang, P. Rao, T. Zhao and X. An .2017. Development and application of EST-SSR markers in *Koelreuteria paniculata* Laxm.using a transcriptomic approach. Biotechnology.16: 45-56.

Young, N. D. and S.D. Tanksley. 1989. RFLP analysis of the size of chromosomal segments retained around the Tm-2 locus of tomato during backcross breeding. Theoretical and Applied Genetics.77:353-359.

Zachariah, T.J., M. Ravindra and M.N. Venugopal. 1998. Quality of cardamom from different accessions. In Mathew, N.M. and Jacob, C.K. (Eds.) Developments in Plantation Crops Research, Allied pub. India, pp. 337-340.

Zaefizadeh, M and R. Goliev. 2009. Diversity and relationships among durum wheat landraces (subconvars) by SRAP and phenotypic marker polymorphism. Research Journal of Biological Sciences. 4(8): 960-966.

Zarini, H. N., H. Jafari, H.D. Ramandi, A.R. Bolandi and M.R. Karimishahri. 2019. A comparative assessment of DNA fingerprinting assays of ISSR and RAPD markers for molecular diversity of Saffron and other *Crocus spp.* in Iran. The Nucleus.62 :39-50.

Zhang, C and S.H. Kim. 2003. Overview of structural genomics: from structure to function. Current Opinion in Chemical Biology.7: 28-32.

Zhang, C., B. Xu, W. Geng, Y. Shen, D. Xuan, Q. Lai,... and C. Yu. 2019. Comparative proteomic analysis of pepper (*Capsicum annuum* L.) seedlings under selenium stress. Peer J. 7. Doi: 10.7717/peerj.8020

Zhang, S., Wei, J. S., and J. Khan. 2014. The Significance of Transcriptome Sequencing in Personalized Cancer Medicine.p.49-64, In G. Dellaire, J.N.Berman and R.J. Arceci, eds. Cancer Genomics From Bench to Personalized Medcine, Academic Press.

Zhang, X. F., B. Chen, L.Y. Zhang, L.L. Zhang, X.H. Chen, H. Zhao and S.S. Geng. 2015. Identification of proteins associated with cytoplasmic male sterility in pepper (*Capsicum annuum* L.). South African Journal of Botany.100:1-6.

Zhang, Z., Y. Zhu, Y. Cao, H. Yu, R. Bai, H. Zhao and L. Wang. 2020. Fine mapping of the male fertility restoration gene CaRf032 in *Capsicum annuum* L. Theoretical and Applied Genetics.133:1177-1187.

Zhao, J., L.C. Davis and R. Verpoorte. 2005. Elicitor signal transduction leading to production of plant secondary metabolites. Biotechnology Advances. 23(4):283-333.

Zhao, S., W.P. Fung-Leung, A. Bittner, K. Ngo and X. Liu. 2014. Comparison of RNA-Seq and microarray in transcriptome profiling of activated T cells. Plos One.9(1).Doi: 10.1371/journal.pone.0078644

Zhao, Y., Y.Liu, Z. Zhang, Y. Cao, H. Yu, W. Ma and L.Wang. 2020. Fine mapping of the major anthracnose resistance QTL AnR GO 5 in *Capsicum chinense* 'PBC932'. BMC Plant Biology. 20:1-8.

Zhao, Z., Y. Guo, W. Zhang, Z. Li, H. Sun, C. Zhang and Q. Guo. 2018. Analysis of Cloning and Expression Characteristics of *Capsicum chinense* Jacq. CcMYB Gene. Agricultural Biotechnology. 7:1-7.

Zhu, X. F., X. Q. Yu and J.L. Zhao. 2019. The complete chloroplast sequence of *Roscoea humeana* (Zingiberaceae): an alpine ginger in the Hengduan Mountains, China. Mitochondrial DNA Part B. 4: 1398-1399.

Zhu, Z., X. Xu, B. Cao, C. Chen, Q. Chen, C. Xiang and J. Lei. 2015. Pyramiding of AtEDT1/HDG11 and Cry2Aa2 into pepper (*Capsicum annuum* L.) enhances drought tolerance and insect resistance without yield decrease. Plant Cell, Tissue and Organ Culture (PCTOC). 120: 919-932.

Zou, X., J.L. Zhao and Z.Q. Zhang.2020.The complete chloroplast genome of *Roscoea tibetica*, an alpine ginger species. Mitochondrial DNA Part B. 5:1878-1879.

Zou, X., Z. Dai, C. Ding, L. Zhang, Y. Zhou and R. Yang. 2011. Relationships among six medicinal species of Curcuma assessed by RAPD markers. Journal of Medicinal Plants Research. 5:1349-1354.

Zych, M., M. Furmanowa, A. Krajewska-Patan, A. Łowicka, M. Dreger and S. Mendlewska. 2005. Micropropagation of *Rhodiola kirilowii* plants using encapsulated axillary buds and callus. Acta Biologica Cracoviensia Series Botanica.47:83-87.

Vavilov, N.I.; Liove, Doris (trans.) (1992). *Origin amd Geography of cultivated plants.* Cambridge University Press. p. xxi. ISBN 978-0521404273.

Annexure

Annexure-I

Table 1: A few spice crop(s) with diploid chromosome numbers

Spice crop(s)	Botanical name	2n Chromosome number
Black pepper	*Piper nigrum*	52
Cardamom	*Eletteria cardamomum*	48 and 52 Mysore type – 50 Malabar type – 48
Ginger	*Zingiber officinale*	22
Turmeric	*Curcuma longa*	63
Vanilla	*Vanilla planifolia*	32
Allspice	*Pimenta dioica*	22
Clove	*Syzygium aromaticum*	22
Cinnamon	*Cinnamomum verum*	24
Nutmeg	*Myristica fragrans*	42
Saffron	*Crocus sativus*	24
Garlic	*Allium sativum*	16
Coriander	*Coriander sativum*	22
Cumin	*Cuminum cyminum*	14
Fenugreek	*Trigonella foenum graecum*	16
Fennel	*Foeniculum vulgare*	22

Annexure-II

Table 2: Variability of quality attributes in black pepper cultivars

Traits	Cultivars
Higher Piperine	Kottanadan, Kumbakodi, Kuthiravally and Nilgiri
Higher Oleoresin	Balankotta, Kaniyakadan and Kumbakody
Higher dry recovery	Kalluvally

Annexure-III

Table 3: Accessions short listed for different breeding traits

Trait	Germplasm accessions
Tolerance/resistance to foot rot disease	IISR Thevam, IISR Shakti
Resistance to Pollu beetle	Acession nos.816 (Neyyatinkara mundi), 841 (Veluthakaniyakadan), 1084 (Cheppukulamundi) and 1114 (Kumbhachola)
Tolerance/resistance to Slow wilt (nematodes)	Pournami, Acession no. 820 (Perumkodi), Hybrid 39 (Irumaniyan x Karimunda)
Tolerance to drought	Karimunda Selection (KS) 69, KS 51 and KS 114, 813 (Ottaplackal-II), 91 (Kalluvally) and 1495 (Kottanadan)
High caryophyllene content	Acession nos.840 (Vattamundi), 971 (Balankotta), 1019 (Vellamunda) and 1022 (Karimunda)

Annexure-IV

Table 4: Commercial quality specifications for turmeric in Europe

Colour	Germany- orange colour, Netherland - yellow colour, UK -deep orange to yellow.
Taste	Powder of peppery odour and bitter taste
Volatiel oil	Germany- 3.5% (min.), Netherland-2% (min.), UK -3.5-4% (min.)
Bulk density	460-500 gm/L in ground powder
Curcumin	3-7% (as per ASTA method 18)
Self life	12 month

Annexure-V

Table 5: Variability for quality traits of ginger cultivar

Sl.No.	Character /trait	Cultivar Accession
1	High Dry recovery	Tura Local (29%), Tura (28%), Thodupuzha (22%), Kuruppampadi (23%), Nadia (22%)
2	High Oleoresin	Rio-de-Janeiro (10.5%), Assam (9.3%), Maran (10%)
3	High Essential oil	Mannanthody (2.2%), Karakkal (2.4%), Ellakallan (2.6%)
4	Least crude fibre	China (3.4%), UP (3.7%), Nadia (3.9%)
5	High Dry ginger yield (t/ ha)	Rio-de-Janeiro (3.27), Thingpruri (2.79), Maran (4.4), Nadia (3.8)
6	High zingiberene and gingerol	Bahrica and Amaravathi
7	Salted Ginger	Acc. Nos. 35 and 37

Annexure-VI

Prominent high yielding and improved varieties of spices available in India

• Black pepper

Panniyur-1, Panniyur-2, Panniyur-3, Panniyur-4, Panniyur-5, Panniyur-6, Panniyur-7, Panniyur-8, Subhakara, Sreekara, Karimunda, Panchami, Pournami, Kottanadan, IISRThevam , IISR Shakti, Vijay etc.

• Ginger

Suprabha, Suruchi, Surabhi, Himagiri, IISR-Varada, IISR-Rejatha, IISR-Mahima, Maran, Rio de Janeiro, Himachal, Valluvanad, Kuruppampady, Aswathy, Karthika, Aathira, Nadia etc.

• Turmeric

CO- 11983, BSR-11986, BSR-21994, Roma, Suroma, Ranga, Rasmi, Rajendra Sonia, Megha Turmeric -1, Pant Peethabh, Suranjana, Duggirala, Alleppey Supreme, IISR Kedaram, Suvarna, Suguna, Sudarshana, IISR Prabha, IISR Prathibha, Kanthi, Sobha, Sona, Varna,Narendra Haldi-1 etc.

• Chilli

K-1, K2, Co-1, , Pusa Jwala, Pusa Sada Bahar, NP 46A, Capsicum – Pusa Deepti, KDC-1, Pant C-1, Pant –C2, Kalyan Sel.1, G1-1962, G2 -1962, G3-1962, Bhagya Laxmi, LCA-305, LCA-334, LCA-235, LCA-353, Punjab Lal, CH-1, CH-2, CH-3, Arka Harita, Arka Lohit, Arka Meghana, Jwalasakhi, Jwalamukhi, Ujjhala, Anugraha, KA-2, Hisar Shakthi, Hisar Vijay etc

• Garlic

Agrifound white, Yamina safed-1, Yamuna safed -2, Yamuna safed-3, Agrifound Parvati, G-323, Pant Lohit-1, Pant Lohit-2, Pusa Sel-10, HG-1, HG-6 etc.

• Cinnamon

IISR-Nithyashree, IISR-Navashree, YCD-1, PPI(C)1, Konkan Tej, Sugandhini, RRL (B), etc.

• Nutmeg

Konkan Sugandha, Konkan Swad, Viswashree, Keralashree etc.

• Tamarind

PKM-1, DTS-1, Prathisthan, No. 263, Yogeswari etc.

- **Curry leaf**

DWA-1, DWA-2, Suvasini etc.

- **Coriander**

Guj. Cor.1, Guj. Cor.2, Co.1, Co.2, Co.3, Rajendra Swati, RCr.41, RCr.20, RCr.435, RCr.436, RCr.446, Sadhana, Swathi, Hisar Sugandh, Hisar Anand, Pant Haritima, Azad Dhania–1 etc.

- **Cumin**

MC-43, Guj Cumin-1, Guj. Cumin-2, Guj. Cumin-3, Guj. Cumin-4, RZ-19, RZ-209, RZ-223 etc.

- **Fennel**

Guj. Fennel-1, Guj. Fennel-2, Co-1, RF-101, RF-125, Hisar Swarup, Azad Sauf-1, Pant Madhurika, Rajendra Sourabha etc.

- **Fenugreek**

Co.1, Co.2, Rajendra Kanti, Rajendra Abha, Hisar Sonali, Hisar Suvarna, Hisar Guj. Neethi, RMt.1, RMt-143, RMt-303, RMt-305, Rajendra Khushba, Pusa Early Bunching etc.

- **Ajowan**

Gujarat Ajowan -1, Pant Ruchika, RPA-68, Ajmer Ajowan -1, Ajmer Ajowan -2, Lamsel-1, Lamsel-2, Rajendra Mani etc.

- **Celery**

RRL-85-1 etc.

- **Dill**

Guj. Dill-1, Guj. Dill-2, RSP-11, Ajmer Dill -1, Ajmer Dill-11 etc.

(Ref: Krishnamurthy, 2018)